Perfume Engineering

Perfume Engineering
Design, Performance & Classification

Miguel A. Teixeira, Oscar Rodríguez, Paula Gomes, Vera Mata, Alírio E. Rodrigues
Laboratory of Separation and Reaction Engineering (LSRE)
Associate Laboratory
Department of Chemical Engineering
Faculty of Engineering of University of Porto
Porto, Portugal

P. Gomes and V. Mata are currently at i-sensis company

AMSTERDAM • BOSTON • HEIDELBERG • LONDON
NEW YORK • OXFORD • PARIS • SAN DIEGO
SAN FRANCISCO • SINGAPORE • SYDNEY • TOKYO
Butterworth-Heinemann is an imprint of Elsevier

Butterworth-Heinemann is an imprint of Elsevier
The Boulevard, Langford Lane, Kidlington, Oxford, OX5 1GB, UK
225 Wyman Street, Waltham, MA 02451, USA

First published 2013

Notices
Knowledge and best practice in this field are constantly changing. As new research and
experience broaden our understanding, changes in research methods, professional practices,
or medical treatment may become necessary.

Practitioners and researchers must always rely on their own experience and knowledge
in evaluating and using any information, methods, compounds, or experiments described
herein. In using such information or methods they should be mindful of their own safety and
the safety of others, including parties for whom they have a professional responsibility.

To the fullest extent of the law, neither the Publisher nor the authors, contributors, or editors,
assume any liability for any injury and/or damage to persons or property as a matter of products
liability, negligence or otherwise, or from any use or operation of any methods, products,
instructions, or ideas contained in the material herein.

British Library Cataloguing-in-Publication Data
A catalogue record for this book is available from the British Library

Library of Congress Cataloging-in-Publication Data
A catalog record for this book is available from the Library of Congress

ISBN: 978-0-08-099399-7

For information on all Butterworth-Heinemann publications
visit our website at store.elsevier.com

This book has been manufactured using Print On Demand technology. Each copy is produced
to order and is limited to black ink. The online version of this book will show color figures
where appropriate.

Working together to grow
libraries in developing countries

www.elsevier.com | www.bookaid.org | www.sabre.org

ELSEVIER   BOOK AID International   Sabre Foundation

CONTENTS

PREFACE

This book tells the story of research on Perfume Engineering in the Laboratory of Separation & Reaction Engineering (LSRE) at the Faculty of Engineering, University of Porto (Portugal). The interest of this story is on the development of a new research line from the scratch, in a field completely new for the laboratory, pushed by the personal interest (and efforts) of the researchers involved.

All starts in 1999, when Alírio E. Rodrigues (AER) proposed a post-doc position to Dr. Vera Mata (VM). VM got a PhD thesis on the characterization of porous media and their application as catalysts' supports. Her personal interests on fragrances and perfumes trigger the jump to the so-called "Perfume Engineering." The aim of her post-doc research was to answer one intriguing question: given a composition of a liquid mixture what do we smell? And can we predict it? From the very beginning, both AER and VM expected that the output of the post-doc was going to be a spin-off company in that area.

To start AER invested in a GC−MS and many commercial perfumes were analyzed. A first attempt to get funding from FCT (our national agency for funding research) failed but VM succeeded in the second trial (SAPIENS 39990/EQU/2001−Design of a perfume using natural resources and clean technologies). She built a homemade Supercritical Fluid Extractor (SCRITICAL) and some essential oils from Portuguese aromatic plants were produced: limonene as top note, geranium oil as intermediate note, rockrose from *Cistus ladaniferus* as base note. In the meantime, Paula Gomes (PG) joined as a PhD student and we had moved to the new campus in 2000. Also an olfactometer was bought and odor thresholds were collected for some fragrances.

The funded project attracted the attention of the media, with interviews in newspapers and television. The engineering contribution became clear with a paper published in the *AIChE Journal* on "Engineering Perfumes" in which the idea of the Perfumery Ternary Diagram (PTD®) was explained. It allowed predicting the odor value for each liquid composition, mapping the various regions of smells in a triangle where the vertices correspond to the three main perfume notes.

The idea of perfume classification (later called the perfumery radar) was shown in a slide at the Product Technology Congress, Groningen, the Netherlands (2004) in the presentation entitled "The science behind perfumes design."

The first PhD thesis on the topic was from PG called "Engineering perfumes" and finished in 2005 with Solke Bruin (the Netherlands) as opponent.

In 2006, VM left the laboratory and created a spin-off company named i-sensis (www.i-sensis.com) dedicated to the development of personalized perfumes and olfactive marketing. She was joined by PG who continued a post-doc in industrial environment for a while.

The second wave continued with Miguel Teixeira (MT) who defended his PhD "Perfume performance and classification: Perfumery Quaternary–Quinary (PQ2D®) and Perfumery Radar" in 2011 with André Chieffi from Procter & Gamble as opponent. During the period of the thesis of MT, Dr. Oscar Rodríguez (OR) joined the group of Product Engineering. The concept of PTD® was extended and the perfumery radar was presented to classify perfumes. Again the impact of that paper in the media (*The Economist*, *Chemical & Engineering News*, among others) was very high.

Under the initiative of VM, a project of "Microencapsulation of perfumes for textile application," funded by Agency of Innovation (AdI) was started, before she left to create i-sensis, in collaboration with CITEVE and a company "A Penteadora" where industrial tests were carried out. That work opened another avenue on microencapsulation inside the Product Engineering group at LSRE. Perfume Engineering is also connected with another research topic on "Valuable chemicals from lignin," started in 1990, aiming at producing vanillin (a perfume ingredient) and syringaldehyde from kraft black liquor which is now gaining visibility with the rise of the biorefinery concept.

Research in Perfume Engineering has attracted the attention of some "big" companies in the area who come to visit the laboratory and start some cooperative work being "surprised" how a group could start meaningful research (from industrial perspective) by his own initiative and ideas. It is true that we were in Academia but with eyes open to the work going on and participating in main conferences as the World Congress on Perfumery (Cannes, 2004) or the CosmInnov–Cosmetic Innovation Days (Orléans, 2010).

Now several young trainees are coming from foreign countries (Poland, France, Brazil) to work in the laboratory, while the third wave started with the study of the matrix effect on perfume performance in relation with personal care and home-care products.

Miguel A. Teixeira
Oscar Rodríguez
Paula Gomes
Vera Mata
Alírio E. Rodrigues
Porto, 2012

CHAPTER 1

A Product Engineering Approach in the Perfume Industry

1.1 THE FLAVOR AND FRAGRANCE MARKET
1.2 FROM THE IDEA TO MARKET: PRODUCT ENGINEERING
REFERENCES

Fragrances are used in a wide variety of daily products like perfumes, cosmetics, toiletries, and household cleaners. The purpose of including fragrances in the formulation of all these different products is to influence consumers, either by enhancing their sensorial properties or simply by signaling the product to be easily recognizable. On their side, consumers are attracted to perfumed products because they are capable of influencing their image, mood, or even their personality. Remarkably, the incorporation of fragrances in products has also the role of improving the evaluations that customers make for the performance of those products: fresh odors are often applied in cleaning products because consumers associate fresh with clean. This bilateral relationship is explained by the power of the sense of olfaction, which surpasses frontiers that other senses cannot reach.

1.1 THE FLAVOR AND FRAGRANCE MARKET

It is no wonder that the business of Flavors and Fragrances (F&F) has become a multibillion dollar market with great economic impact all over the world. Currently, it includes two different fields of operation: (i) the production of raw materials, either extracted from natural sources or synthesized in the laboratory and (ii) the formulation of flavor or fragrances' blends. Of course, there are many other industries operating in diversified areas of expertise that are also closely related to the F&F business like packaging, marketing, or retail chain companies. In its entirety, the global market of F&F is large and has been continuously growing at good rates over the last decade, featuring an average growth of more than 5% per year. In spite of the economic

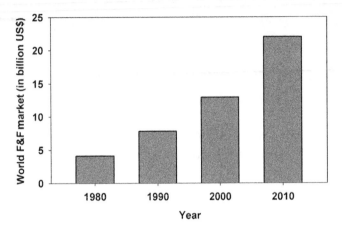

Fig. 1.1. Evolution of the global F&F market in billion US$ for the period 1980–2010. The average annual growth rate exceeds nearly five times the corresponding world population growth, which is estimated at 1.1%. https://www.cia.gov/library/publications/the-world-factbook/geos/xx.html.

and financial crisis of 2008, astonishingly the F&F worldwide market has more than doubled in the past 15 years, from US$9.6 billion in 1995 to US$22 billion in 2010 as shown in Fig. 1.1 (Leffingwell and Leffingwell, 2012).

Despite the magnitude of this market and the large number of companies operating within its frontiers, the fact is that it is mainly controlled by a restricted group of 10 companies. This is so, because the dynamics of the F&F industry follows the trend of other industries where several mergers, acquisitions or market expansions from the most representative companies (Ziegler, 2007) often occur. The "big fish" companies dominate a total of 76% of the market share, and the top five companies (Givaudan, Firmenich, International Flavors & Fragrances—IFF, Symrise, Takasago) account for more than 60% alone. A summary of such ranking of F&F companies is given in Table 1.1. Consequently, it can be said that this is a closed and strong market that still remains ruled by a small group of companies, although there are hundreds of other small companies operating within it. Perhaps it is one of the reasons behind the secrecy in this industry. To a layman or a regular consumer of perfumes, these companies will be mostly unfamiliar since they do not appear on the shelves of perfume shops or on the packaging of perfumes. And the publicity they receive in the media is rare or even nonexistent. Probably, companies' names such as Christian Dior,

Table 1.1. World Market Share for F&F Business per Company in 2010 (Values in Million Dollars)			
Rank	Company	US$	Market Share (%)
1	Givaudan	4538	20.6
2	Firmenich	3319	15.1
3	IFF	2623	11.9
4	Symrise	2107	9.6
5	Takasago	1416	6.4
6	Mane SA	643	2.9
7	Sensient Flavors	583	2.6
8	T. Hasegawa	557	2.5
9	Robertet SA	485	2.2
10	Frutarom	451	2.1
Top 10	–	16,722	76.0
All others	–	5278	24.0
Total market	–	21,999	100.0

Dolce & Gabbana, Estée Lauder, or Hugo Boss will sound more familiar. However, the fact is that the vast majority of these companies (which work as brand managers) neither do not produce the fragrances themselves nor have perfumers in their staff. In fact, what happens is that when these brand companies spot a gap in their portfolio, then they brief the fragrance houses which will develop the perfume. Thus, companies like Givaudan, Firmenich, or IFF are not only global suppliers of fragrances and flavors (including raw materials and active ingredients for perfumes, cosmetics, and foods) but also manufacturers of perfumes and fragranced products.

Another relevant aspect to be highlighted is the comparison between the geographical distribution of main F&F companies, as depicted in Fig. 1.2, with the number of sales. Geographically, the consumption of products containing either flavors or fragrances is asymmetric and in line with the socioeconomic development of countries worldwide. Thus, North America and central Europe are the gross consumers of such products, contributing to more than 50% of the worldwide consumption (Leffingwell and Leffingwell, 2012). They are followed by growing markets like Japan and China, as presented in Fig. 1.3.

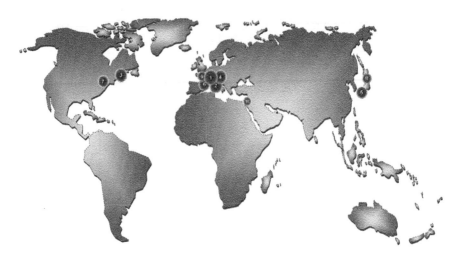

Fig. 1.2. Geographical distribution of the top 10 F&F companies in the world in 2009: 1. Givaudan; 2. Firmenich; 3. IFF; 4. Symrise; 5. Takasago; 6. Mane SA; 7. Sensient Flavors; 8. T. Hasegawa; 9. Robertet SA; 10. Frutarom. Adapted with permission from Teixeira (2011). © 2011, M.A. Teixeira.

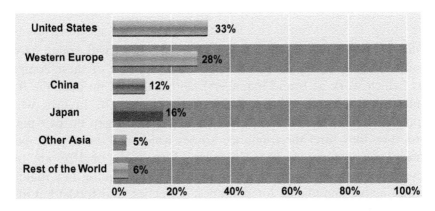

Fig. 1.3. World consumption of F&F products in 2010. Adapted with permission from Teixeira (2011). © 2011, M.A. Teixeira.

After analyzing the F&F market, it is now time to unravel what is inside F&F companies and what products are derived from them. As in analogy with the market, the core business of the F&F industry can be divided into two main groups: fragrances and flavors. While fragrances are odorous organic chemicals that are used in perfumed products, flavors are intended for the flavoring of foods and beverages. Most of what we perceive from flavors is due to the olfactory perception as well, and so these two fields are closely linked. Hence, their classification depends whether they are mainly perceived by the sense of olfaction or gustation. In simple words, whether they are applied in perfumed or food products, respectively. However, one should not forget that much of what we perceive by our taste is actually influenced by what our nose perceives at the same time. Having said that, sales of fragrances and flavors are nearly equivalent, although flavors have been gaining little ground during the last decade. This increasing need for flavors can be explained by the shift in human lifestyle and philosophy: at the turn of the twenty-first century, the trend in consumers started to be oriented toward healthy, fitness, and diet products (moved by the slogan "you are what you eat"), and thus pushing the industry toward such products. Within the markets for fragrances and flavors, it is possible to have trading of either raw materials (natural, natural–identical, or synthetic) or blends of them. For their part, blends constitute added-value products with more market weight. In the field of flavors, aromas are essentially designed for beverages, bakery, savory, and meat products. In the case of fragrances, the largest application goes to functional perfumes designed for everyday use, which should not to be confused with fine perfumery (luxury products). Examples of such products are soaps, detergents, toiletries, and household cleaners. However, the development of fine fragrances, which are more expensive and have higher added value, is also significant (reaching 21% of the fragrance market).

1.2 FROM THE IDEA TO MARKET: PRODUCT ENGINEERING

The previous numbers give an idea of the pressure in the F&F industry to produce more and more profitable fragranced products. In fact, as happens in many other businesses, profit is the goal. Thus, if the ideal target would be to create a new and unique piece of art, it is also true that fragrance houses are often asked to develop a fragrance that is appealing to all types of people and will return millions of dollars (Burr, 2008). That explains the importance of product development departments in this industry. Just as an example, and considering the exclusive fine fragrance market for 2009, over 1500 new fine fragrances were released only during that year (compared with less than 50 new ones, 20 years ago). This quest for novel products is a recent trend that has been growing over the last decade to fulfill consumers' needs and expectations. Currently, it is the market that pulls companies (and their product development teams) to overcome barriers and challenges for the creation of new products. This issue has been explored in a multitude of companies for many years, following a more or less empirical methodology. Nowadays, it is a subject of its own: what is now called Product Engineering, product development, or product design (Charpentier, 1997; Ulrich and Eppinger, 2000; Cussler and Moggridge, 2001; Mata et al., 2004; Wei, 2007; Wesselingh et al., 2007; Cussler et al., 2010). Within the chemical business it is called chemical product design. Books addressing this topic started to show up in the beginning of the 1990s but until now they are less than a couple dozens. The most successful so far is probably that of Karl T. Ulrich and Steven D. Eppinger which is already at its fifth edition (Ulrich and Eppinger, 2011). In terms of scientific papers, the numbers are larger, following a similar trend: started with few in the beginning of the 1990s, but since then they have been increasing as shown in Fig. 1.4. Probably, it is more relevant to the fact that product design and development has become a discipline of many curricula of university courses. As stated in the 1990s by J. A. Wesselingh, S. Kiil, and M. E. Vigild, who wrote a book and give lectures on this topic: *large changes were coming in the chemical industry, and that they should be looking at higher-value (structured) products.*

We consider that Product Engineering can be seen as a stepwise methodology: it starts with the identification of market needs, then their

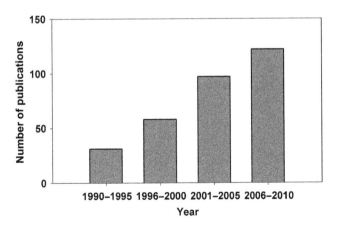

Fig. 1.4. Track record of scientific papers with "Product Engineering" appearing in topics published in international journals from 1990 to 2010. Scopus, "Product Engineering" in title, abstract, and key words, Date of consultation February, 2012.

translation into product specifications (physicochemical properties that can be measured) which will help generating ideas and, finally, ends up with the selection of the best ones for the manufacture of the product. Throughout all this process, it must encompass other relevant topics like economic evaluations, risk assessment, project management, and sustainability. At a glance, it defines the "how," "when," and "where" a new product should be developed and launched into a market. This perspective is shared by several authors in the literature as well. Cussler and Moggridge (2001) consider that Product Engineering *emphasizes decisions made before those of chemical process design, a more familiar topic* for chemical engineers. Moreover, according to Ulrich and Eppinger (2011), *Product development is the set of activities beginning with the perception of the market opportunity and ending in the production, sale, and delivery of a product.* In our opinion, the picture of Product Engineering (or product design and development) drawn by the different authors looks very similar because they share the same fundamentals. Of course, small differences arise as different authors give emphasis to different aspects of the development process. But the industry itself is not away from these approaches: a similar stepwise framework, widely used, is the Stage-Gate™ Product Development Process (SGPDP). It is based on what the industry project teams do better and applies those product design strategies based on decision analysis to the development of novel products (Cooper, 2001; Seider et al., 2010). Having said that, should we be able to apply the fundamentals of Product Engineering to the F&F business? F&F companies are no exception on this matter as they are consumer products companies (e.g., Johnson & Johnson or Procter & Gamble, among many others).

In fact, Product Engineering has everything to do with the countless fragranced products we contact everyday. These may have different properties and functions (e.g., detergents, shampoos, creams, candles, and perfumes) but the incorporation of fragrance ingredients is intended to instill a pleasant and harmonious odor to the product in which it has been incorporated. However, developing a fragrance is a complex and long process that starts in the brief, a brainstorming meeting of the different people involved in its formulation. At that point, many characteristics of the perfume are defined: who the perfume should appeal to and why, what the scent should say to the contractor (if one exists) or, ultimately, to the consumers, what forms the fragrance will take (e.g., spray, *parfum, eau the toillete*, after-shave, and soap), where and for how long the product will be sold (e.g., Europe or America, one or two years), among many other questions. The answers collected from the brief will define the customers' *needs*, how will they influence the *ideas* to be generated, and the *selection* of fragrance ingredients from the F&F company for the *manufacture* of the target product. In short, the brief is a creative and detailed definition of what the new fragrance is supposed to be. The parallelism with the Product Engineering framework, based on the *needs—ideas—selection—manufacture*, is evident. Today, the "secret formula" of a perfume is idealized by perfumers, experts with a high level of experience in the perception of scents (detection and recognition) as well as in the art of creating accords and perfumes. A perfumer can be viewed as a *sommelier* who has no difficulty in discriminating between aromas and notes in the bouquet of a red wine. Perfumers select the ingredients to be used and define their proportions based on their expertise. Their selection depends on the theme of the fragrance house they work for or the consumer desire, and may include fragrance ingredients from different sources (natural, synthetic or natural—identical fragrance ingredients, or essential oils). From that point, the mixing of the ingredients can be performed in assembly lines controlled by laboratory technicians. Nevertheless, in the end, the final formulation will have to rest and age in tanks for several weeks in what are called the maceration and maturation processes (Lopez-Nogueroles et al., 2010).

However, different types of fragrances are formulated for different types of applications or end-use products and not only for the exclusive fine fragrance market. For example, the incorporation of a fragrance in a dishwasher detergent is expected to produce a fresh scent when opening the washing machine and at the same time it is engineered not to leave residual odors on the surface of plates (something complex to achieve from the surfactants perspective). Furthermore, it is known that a fragrance, when incorporated in a product base (e.g., glycerin, which is the main base for soap), may produce an unexpected behavior in terms of perceived odor, stability, or color (among others). These phenomena cannot be solved simply from art and common sense. It requires knowledge from different scientific fields which should be combined in order to understand the interactions within a product. That is what Product Engineering is about. In fact, companies producing fragranced products have top quality scientists in their staff and use sophisticated analytical tools in their laboratories. Still, the design of fragrances remains mostly empirical, based on the experience and know-how of perfumers. Carles (Carles, 1962) stated 50 years ago that most of the greatest and commercially successful perfume creations were produced by serendipity, sometimes to the unfeigned surprise of their authors. But although the technology and the knowledge has evolved, perfume creation is still an art, before being a well-defined science. For example, the perfume "Bois de Paradis" launched in 2005 and developed by Michel Roudnitska took more than two years and nearly 300 formulation trials until he got the desired scent. Another example is the "Tommygirl" fragrance designed by Calice Becker which took change for 1100 iterations until arriving to the market (Wolfson, 2005). Such facts make us realize that it is undisputable that, from the economical perspective of the business, this is undesirable. However, the development of a perfume is not like the manufacture of a car. Although the latter may take two to four years to reach the market (from generating the original idea to its materialization), it has high technological features under its hood that are improved year after year. On the other hand, a new fragrance may seem to be "simply" a mixture of N fragrance ingredients, solvents, stabilizers, antioxidants, UV filters, coloring agents, among others (despite it is also tailored to the client wishes). Once on the assembly line, a car takes somewhere from one to two days until it is finished. A perfume, for its part, may take weeks or even months due to the need for maceration and maturation processes (Calkin and Jellinek, 1994; Curtis and Williams, 1994). Consequently there is no doubt that a significant part of the knowledge on F&F remains under wraps, mainly due to the powerful and closed market of big companies, which consequently leads to secrecy. Nevertheless, the importance of enhancing scientific knowledge in an area like F&F with direct application in the development of products for the consumer is especially relevant for two reasons: (i) the number of fragrance ingredients is in the order of thousands (and increasing) which makes an almost infinite number of possible formulations for different applications and, thus, a corresponding number of trial-and-error evaluations and (ii) it is still dependent on the high skills and expertise of the perfumers.

Thus, the application of Product Engineering to F&F has the crucial objective of implementing technical and scientific knowledge into a so far empiric and experimental area. That is what we call Perfume Engineering: a research line developed in our laboratory for more than a decade and combining different scientific fields:

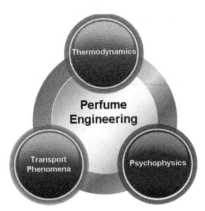

Fig. 1.5. Our vision of what is Perfume Engineering: the integration of concepts from different fields (Thermodynamics, Transport Phenomena, and Psychophysics) in order to improve the design and development of new perfumed products. Adapted with permission from Teixeira (2011). © 2011, M.A. Teixeira.

Thermodynamics, Transport Phenomena, and Psychophysics into a common product–fragrances. A graphic representation of this idea is shown in Fig. 1.5. Perfume Engineering, as we will see it, aims at reducing the time needed for the design of perfumes and fragranced products in the preformulation step and the reduction of the consumption of raw materials. Altogether, that will contribute to decrease the production costs. How we propose to do this will be explored throughout this book.

At this point, it is important to summarize the perspective of the authors for the process of odor perception of perfumed products. We consider that the starting point of this phenomenon is at the bulk of the liquid mixture of fragrance ingredients. That means the selection and composition of fragrances, which are the variables the perfumer can control, and the chemistry within them. From that mixture, fragrance ingredients are being released into the air above it, and later are perceived with some odor character and intensity. This sequence of processes can be depicted in the four steps shown in Fig. 1.6.

In this way, the starting point are the fragrance ingredients and solvents mixed within a homogeneous liquid solution of known molar compositions (x_i) which can be sprayed on the skin or clothes by customers; after application of the product, the process of perception by customers proceeds as follows: (i) the different fragrance chemicals

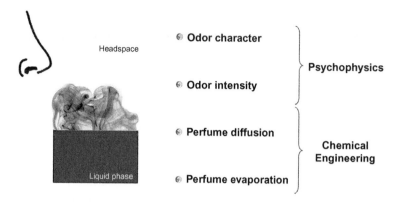

Fig. 1.6. *The process for perception of a perfumed product, divided into four steps: the evaporation of the fragrances used in the formulation, their diffusion through the surrounding air, and when the olfactory system is reached, the perception of the odor intensity, and the character of the mixture. The first two steps belong to the Chemical Engineering area of knowledge, while the two other steps belong to Psychophysics.* Adapted with permission from Teixeira (2011). © 2011, M.A. Teixeira.

begin to evaporate into the headspace, although at different rates depending on their volatility, composition, and molecular interactions; (ii) subsequently, the gas odorant molecules will diffuse through the surrounding air over time and distance; and (iii) finally, at a given time and distance, some of the fragrance molecules will eventually reach the nose of the customer who will perceive the odorants with a certain intensity and character.

Our approach combines different scientific tools within Perfume Engineering: steps (i) and (ii) are within Chemical Engineering and, more specifically, deal with Thermodynamics and Transport Phenomena. On the other hand, step (iii) makes part of Psychophysics, a science within psychology that deals with the mathematical relationships between perceived sensation and its stimulus magnitude. Altogether, different models can be used in each step for the description of the corresponding phenomenon (the perfumed product can be a liquid, a solid, a gel, and fragrance propagation through air can be governed by molecular diffusion or convection). This will allow mapping the perceived odor from a mixture of fragrance ingredients. In the next chapters, this model for the perception of odors elicited from perfumed products will be explained, presenting some details about its validation and application to develop other tools for the F&F industry.

From Fig. 1.6, we see that in the four steps within our perception model, it is possible to take some assumptions and use different

Table 1.2. Different Models for Each of the Four Steps Within Our Methodology for Odor Perception

Perfume Evaporation	Perfume Diffusion	Odor Intensity	Odor Character
Ideal	Fick Law	Odor Value	Stronger Component
		Power Law	Vectorial
Nonideal	Maxwell–Stefan	Weber–Fechner	U and UPL2
		Beidler	Additivity

Further details on the definition and application of these models will be discussed throughout the chapters.

approaches for each of the underlying processes, as summarily presented in Table 1.2. Throughout Chapter 2, we will address fragrance evaporation and perceived odor intensity, by showing the relevance of the assumptions made in each one, together with the challenges resulting from them. Then, in Chapter 3, we will evaluate the performance of fragrances by including the diffusion of volatiles in air through time and space. In Chapter 4, we will address the quality of the perceived odor by predicting the classification of perfumes into olfactive families. Finally, in Chapter 5, we will post some new trends and hot topics that are expected to lead the way in the fragrance business in the near future.

REFERENCES

Burr, C., The Perfect Scent: A Year Inside the Perfume Industry in Paris and New York, Henry Holt and Co., 2008.

Calkin, R., Jellinek, S., Perfumery: Practice and Principles, New York, NY, John Wiley & Sons, 1994.

Carles, J., A method of creation in perfumery, *Soap Perfumery Cosmet.*, 35, 328–335, 1962.

Charpentier, J. C., Process Engineering and Product Engineering, *Chemical Engineering Science*, 52, (18): 3–4, 1997.

Cooper, R. G., Winning at New Products: Accelerating the Process from Idea to Launch, New York, NY, Perseus Publishing, 2001.

Curtis, T., Williams, D. G., Introduction to Perfumery, New York, NY, Ellis Horwood, 1994.

Cussler, E. C., Moggridge, G. D., Chemical Product Design, Cambridge, MA, Cambridge University Press, 2001.

Cussler, E. L., Wagner, A., Marchal-Heussler, L., Designing Chemical Products Requires More Knowledge of Perception, *AIChE Journal*, 56, (2): 283–288, 2010.

Leffingwell, J. C., Leffingwell, D., Flavor & Fragrance Industry Leaders, 2012, http://www.leffingwell.com/top_10.htm, Accessed in October 2012.

Lopez-Nogueroles, M., Chisvert, A., Salvador, A., A Chromatochemometric Approach for Evaluating and Selecting the Perfume Maceration Time, *Journal of Chromatography A*, 1217, (18): 3150–3160, 2010.

Mata, V. G., Gomes, P. B., Rodrigues, A. E., Science Behind Perfume Design, Second European Symposium on Product Technology (Product Design and Technology), Groningen, The Netherlands, 2004.

Seider, W. D., Seader, J. D., Lewin, D. R., Widagdo, S., Product and Process Design Principles: Synthesis, Analysis and Evaluation, Asia, John Wiley & Sons, Inc., 2010.

Teixeira, M. A., Perfume Performance and Classification: Perfumery Quaternary-Quinary Diagram (PQ2D®) and Perfumery Radar. Department of Chemical Engineering, Faculty of Engineering of University of Porto, PhD Thesis, 2011.

Ulrich, K. T., Eppinger, S. D., Product Design and Development, 2nd edition, New York, NY, McGraw-Hill, 2000.

Ulrich, K. T., Eppinger, S. D., Product Design and Development, 5th edition, New York, NY, McGraw-Hill, 2011.

Wei, J., Product Engineering. Molecular Structure and Properties, New York, NY, Oxford University Press, 2007.

Wesselingh, J. A., Kiil, S., Vigild, M. E., Design and Development of Biological, Chemical, Food and Pharmaceutical Products, Chichester, Wiley, 2007.

Wolfson, W., In the Fragrance Molecule Smells Business, the Right Like Money, *Chemistry & Biology*, 12, (8): 857−858, 2005.

Ziegler, H., Flavourings: Production, Composition, Applications, Regulations, Weinheim, Wiley-VCH Verlag GmbH & Co., 2007.

CHAPTER 2

Design of Perfumes

In this chapter, the Perfumery Ternary Diagram (PTD®) methodology is presented as a tool for the prediction and mapping of the odor character of ternary to quaternary mixtures of fragrance ingredients (Mata et al., 2004, 2005a,b,c; Mata and Rodrigues, 2006; Gomes et al., 2008). To present this methodology, the effect of different base notes in simple fragrance mixtures of the type (top note + middle note + base note and/or + solvent) will be studied to illustrate the potential of this tool. In this study, the selected base notes were three ingredients commonly used in perfumery: vanillin, tonalide, and galaxolide. Moreover, as a proof of concept, the PTD® predictive tool is experimentally validated using headspace gas chromatography (GC) techniques.

The PTD® is further extended to the Perfumery Quaternary–Quinary Diagram (PQ2D®) methodology. This novel tool can be applied to quaternary and quinary fragrance systems for the prediction of their headspace odor character and odor intensity using three-dimensional tetrahedric diagrams. The effect of the base note on the perceived odor space will be evaluated with some examples of quaternary mixtures of the type (limonene + geraniol + base note + ethanol). The PQ2Ds of these perfumery systems showed different headspace odor qualities, depending on the base note, thus confirming some perfumery evidences. The PQ2D$^{®}$ methodology is also applied to quinary systems (limonene + geraniol + vanillin + tonalide + ethanol) to evaluate the effect of different concentrations of the fixative (tonalide). Other quinary systems are also presented to evaluate the influence of water on perfume formulation. The perceived odor in the headspace is then compared with that of the corresponding quaternary mixtures (concentrated mixture, without water). Finally, a last effort was put on the graphical representation capabilities of the PQ2D$^{®}$ methodology which was extended to incorporate octonary mixtures with two top, middle, and base notes and two solvents.

2.1 THE PERFUMERY TERNARY DIAGRAM

In the beginning of the research line on Perfume Engineering at our laboratory, one important question that we wanted to answer was raised: when several fragrance ingredients are mixed together (like in a perfume), what will its perceived smell be like? In fact, when holding soap in hand, we may question ourselves why this product smells so good? We know that fragranced products are, as we said before, the result of the vast knowledge and expertise of perfumers. Nevertheless, in a world that avidly seeks for answers to every new scientific fact, it makes sense to question the way a perfumed product is perceived by us. Indeed, these are big questions for the perfume business as well: a perfume is a combination of many fragrance ingredients, often in the order of 50–100. Very few perfumes come to the market with less than 30 fragrance chemicals in their formulation (Angel from Thierry Mugler is a rare example). Yet, fragrances for cleaning products or toiletries are often much simpler in composition. Of course, apart from the fragrances, other types of components are also included in the formulation (solvents, stabilizers, colorants, and UV filters), but fragrance ingredients are the responsible for the smell, which is the core of the product.

The path to answer the above questions was sketched at the end of Chapter 1 with the methodology we have developed to model and predict odor perception. But, which would be the best way to show the perceived odor to the people involved in the formulation of a perfume? It was important to develop a tool that would be able to present the answer in a suitable and understandable way — that was the PTD®. The PTD® methodology started to be outlined in the year 2003 and came to light in an article published in 2005 (Mata et al., 2005c). In a glimpse, the PTD® is a graphical software tool for the prediction of dominant odors as perceived by humans for any possible ternary mixture of fragrance ingredients. The development of this idea started with the symmetry between the pyramidal perfume structure proposed by Carles (1962) and a typical ternary diagram, often used in engineering. Such symmetry is shown in Fig. 2.1. Ternary diagrams are commonly used in different fields of Chemical Engineering (e.g., Thermodynamics for phase equilibria, process separation design). The diagram represents the compositions of three components as defined by its edges, so each vertex stands for a pure component (100%), while the opposite side means the lack of that component (0%) (Perry, 1997).

In the last century, the famous perfumer Carles (1962) stated that a well-structured perfume must be the combination of top, middle, and base notes. Top notes are the most volatile fragrances, which are perceived right after the application of the perfume and can last for several minutes. Middle notes are less volatile, and so should be more strongly perceived after top notes faded away, lasting for few hours. Finally, base notes are the less volatile, being more strongly perceived after the middle notes have disappeared and lasting many hours or even days. Carles ordered these notes in a pyramid with different layers as in the top-left part of Fig. 2.1, where each section also represents the "recommended" proportions in their combination: top (15−25%), middle (20−40%), and base, (45−55%). He postulated that these proportions were responsible for the tenacity of the perfume, in other words, the balanced evolution of the odor during evaporation. Accordingly, the PTD® mixes both concepts, placing a top, a middle, and a base note at the vertices of the triangle. The points inside the triangle represent all possible ternary mixtures. In this way, it is feasible to map in the diagram the regions (composition ranges) where each fragrance ingredient or note have the dominant odor among all of them, if we are able to calculate the odor intensity for each of these mixtures. It should be noted,

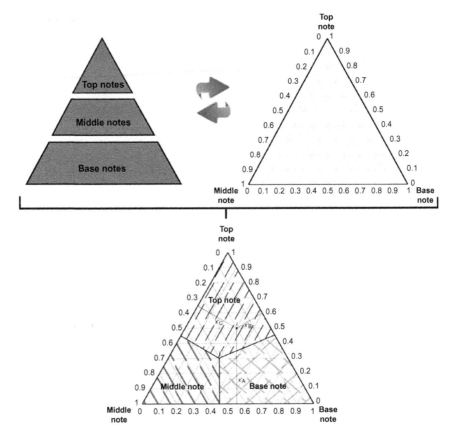

Fig. 2.1. Relationship between the pyramid structure of a perfume (top left) and the engineering ternary diagrams (top right), resulting in the PTD®. In the PTD® different zones represent the dominant perceived odor from the ternary mixture. Adapted with permission from Teixeira (2011). © 2011, M.A. Teixeira.

at this point, that depending on the selection of components and their solubility, there may be some compositions that do not satisfy a homogeneous liquid solution (a requisite for a perfume).

Each single point inside the ternary diagram represents a unique ternary mixture of the type A + B + C (where the sum of their fractions must be equal to unity). Consequently, it is possible to study perfume formulations of the type (top + middle + base notes), as expected for a perfume concentrate.

2.1.1 Odor Intensity Model

At this point, we are interested in predicting or calculating the perceived odor character of each single point (equivalent to each single mixture) within the ternary diagram. In order to do so, an odor intensity model is required to calculate the perceived magnitude of odors from the vapor concentrations. There are several such models derived from Psychophysics for single components like the Weber–Fechner Law, Stevens' Power Law, Odor Value (OV) concept, Beidler model, Laffort model, among others (Müller et al., 1993; Cain et al., 1995; Teixeira et al., 2010). In this book, we will adopt the concept of OV and the Power Law (Ψ) (Stevens, 1957; Calkin and Jellinek, 1994). Some authors use the term Odor Activity Value (OAV) which is equivalent to the OV. For its part, the OV is a quantitative parameter that defines the odor strength (or odor intensity) of an odorant species i as the ratio between its concentration in the headspace, C_i^g, and its odor detection threshold in air, ODT_i (Calkin and Jellinek, 1994; Zwislocki, 2009; Ohloff et al., 2012). The term headspace is used to represent the gas phase (air) above a mixture of fragrance raw materials (Calkin and Jellinek, 1994; Curtis and Williams, 1994; Ohloff et al., 2012). The equation is as follows:

$$OV_i = \frac{C_i^g}{ODT_i} \tag{2.1}$$

In the case of the Power Law, the relationship is raised to power exponent and defined as:

$$\Psi_i = \left(\frac{C_i^g}{ODT_i}\right)^{n_i} \tag{2.2}$$

where C_i^g is the concentration of the odorant in the gas phase and ODT_i is its corresponding odor detection threshold in air (both using units of mass or mol per volume). The parameter n_i is the power law exponent for each odorant.

The ODT represents, in simple terms, the minimum concentration of an odorant that can be perceived by the human nose. A more systematic definition is reported by the ASTM (Method E 679-91) which defines the ODT as *the concentration of an odorous compound at which the physiological effect elicits a response 50% of the time* (Mayer et al., 2005). There are different types of odor thresholds (detection, recognition, terminal, difference) which we will not discuss in detail here, but we encourage the interested reader to see the references in the literature (Hau and Connell, 1998; Cain and Schmidt, 2009; Teixeira et al., 2011a,b). Nonetheless, the most important ones are the detection

(ODT$_i$) and the recognition (ORT$_i$) thresholds (the latter represents the minimum concentration for the recognition of the smelled odorant). We consider the ODT far more reliable than the ORT and thus, the former is selected to be used in our studies. Extensive compilations of odor threshold values can be found in the literature for odorant substances in air, water, and other media (Patte et al., 1975; Calkin and Jellinek, 1994; Nagata, 2003; van Gemert, 2003; Leffingwell and Leffingwell, 2012). However, care must be taken when using odor thresholds, especially concerning to the experimental technique, methods, and equipments used, room conditions and panel composition, and statistical data treatment. As a result, despite the vastness of odor threshold data available in the literature, their experimental measurement is difficult, time consuming, and labor intensive. It is no surprise that it inherently presents a large variability between laboratories and over the years, mainly because this parameter has a large physiological variability (Cain and Schmidt, 2009; Teixeira et al., 2011a). Consequently, experimental threshold data should be used with a critical eye. Nevertheless, their application and relevance is widely recognized and odor thresholds are applied in a multitude of fields (AIHA, 1989; Cain and Schmidt, 2009).

Recalling Eq. (2.1), it results that the OV is a dimensionless parameter (because both variables must have the same concentration units despite their selection is arbitrary, e.g., g/m^3). It defines the potency of an odorant in a very simple way. Additionally, the above definition shows a linear relationship between the stimulus magnitude and its perceived sensation. Hence, if the odorant concentration in the headspace is multiplied by a factor of two, the magnitude of the OV will double. In the case of Eq. (2.2), it is seen that the Stevens Law presents a linear relationship for low odorant concentrations (near threshold and when the exponent equals unity) but then approximates to a *plateau* at higher concentrations (mostly because odorant exponents are generally below unity). This equation can be plotted in a log–log scale to obtain a linear relationship between a stimulus and its perceived sensation, where the slope equals the exponent n and the intercept is a function of the odor threshold, as presented in Fig. 2.2.

Although the OV may be a rough approximation to the real phenomenon of odor quantification, we will also see that when incorporated in the PTD$^{\circledR}$ it is able to give a good perspective for the

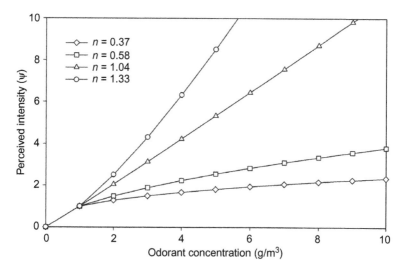

Fig. 2.2. *Relationship between the perceived intensity and the odorant concentration for different exponents using the Stevens' Power Law (limonene, n = 0.37, ethanol, n = 0.58, butylamine, n = 1.04, sucrose for taste, n = 1.33). In olfaction, the majority of the odorants generate power functions with exponents smaller than unity, and for the compilation of 213 data from Devos et al. (2002) the most frequent value was equal to 0.35.* Adapted with permission from Teixeira (2011). © 2011, M.A. Teixeira.

qualitative odor perception. From simple inspection of Eq. (2.1), it is clear that only the volatiles with an OV higher than unity will be detected by the human nose (suprathreshold), while values below unity will be too diluted to be detected (subthreshold).

2.1.2 Vapor–Liquid Equilibrium

Yet, as we know, a perfume, no matter how simple it is, will always be a liquid mixture of N fragrance ingredients. Consequently, the smell we perceive above the liquid is not a single odorant only, rather a complex mixture of them with different concentrations and odor intensities. In order to model the molecular interactions between those and to evaluate that behavior, some basic Thermodynamics can be used. The composition of the different fragrant chemicals in the gas phase above the liquid (headspace) can be calculated from a modified Raoult's Law for vapor–liquid equilibria (VLE), as presented in Eq. (2.3).

$$y_i \phi_i P = x_i \gamma_i P_i^{sat} \tag{2.3}$$

where y_i and x_i are the vapor and liquid mole fractions of component i, while ϕ_i and γ_i are the vapor and liquid activity coefficients of component i, respectively. P represents the total pressure and P_i^{sat} is the saturation pressure of pure component i. At atmospheric pressure, ideal gas behavior can be assumed so that Eq. (2.3) can be simplified by considering $\phi_i = 1$. Consequently, the concentration of odorant species in the headspace (C_i^g) can be calculated from Eq. (2.4).

$$C_i^g = \frac{y_i M_i P}{RT} = \gamma_i x_i \frac{P_i^{sat} M_i}{RT} \tag{2.4}$$

where M_i is the molecular mass of component i, R is the universal gas constant, and T is the absolute temperature. Combining the definition of OV, previously presented in Eq. (2.1), with Eq. (2.4), it is possible to express the OV of each fragrant component by the following equation:

$$OV_i = \gamma_i x_i \left(\frac{P_i^{sat} M_i}{ODT_i}\right)\left(\frac{1}{RT}\right) \tag{2.5}$$

It should be highlighted that the composition in the liquid mixture (x_i) appears in Eq. (2.5) and that is the variable that perfumers can control when designing their fragrances. As a result, it is possible to calculate the odor intensity of fragrance chemicals from pure component data (composition in the liquid phase, molecular weight, saturated vapor pressure, and odor detection threshold) together with the activity coefficient. However, some implications regarding activity coefficients must be clarified. First, let us attempt to describe what it represents: it is a "virtual" measure of the molecular interactions occurring between molecules within a mixture (between fragrance ingredients themselves and between those and solvents as well). In simple terms, it accounts for deviations of the liquid phase from ideal behavior, reflecting the affinity of each molecule with its surrounding medium (Poling et al., 2004). Recalling Raoult's Law, both gas and liquid phases are assumed as ideal, so $\phi_i = \gamma_i = 1$, as shown in Eq. (2.6).

$$y_i P = x_i P_i^{sat} \tag{2.6}$$

Comparison of Eqs (2.3) and (2.6) shows that γ_i can be understood as a measure of the tendency of a fragrance i to be "retained in" or "pushed out" of the perfume. Thus, if $\gamma_i > 1$, the fragrant component i will be more pushed out from the solution into the gas phase (and so the headspace concentration will be higher than for an ideal solution), while $\gamma_i < 1$ expresses that it will have more tendency to stay in the liquid solution due to a higher affinity with the surrounding medium (and consequently lower concentrations of that odorant species will be found in the headspace) (Mata et al., 2005b).

Activity coefficients can be calculated from rigorous experimental VLE data, or from thermodynamic models such as NRTL (Renon and Prausnitz, 1968), or UNIQUAC (Abrams and Prausnitz, 1975). They can also be predicted using methods like UNIFAC (Fredenslund et al., 1975; Poling et al., 2004), ASOG (Tochigi et al., 1990), COSMO-RS (Klamt, 2005), or Molecular Simulation techniques (Panagiotopoulos, 1992; Gubbins, 1994). Here, we will use, in a first approach, the original UNIFAC method for prediction of the VLE and the activity coefficients of fragrance components. The need for activity coefficients lays in the nonideality of fragrance mixtures: fragrance molecules often involve the presence of many different functional groups, and consequently many different types of interactions are in play. Neglecting the nonidealities (i.e., direct use of Eq. (2.6)) means that a linear relationship between liquid and gas phase concentrations is assumed for the whole concentration range. Such linearity can be misrepresentative of the reality in many cases.

2.1.3 Odor Quality Model

Going back to Eq. (2.5), it relates the OV of a fragrance (which is a measure of its odor intensity) to its concentration in a liquid mixture. At this point, we just need to model the character of the olfactory perception of all odorants present in the mixture from their odor intensities. We can achieve that using the Stronger Component (SC) model: it is an approximation to the human odor recognition, which defines that within a mixture of odorants the one that will be more strongly perceived and recognized is the one having the highest OV (Laffort and Dravnieks, 1982; Cain et al., 1995):

$$OV_{mix} = \max\{OV_i\}, \quad i = 1, \ldots, N \qquad (2.7)$$

The use of the SC model to account for the odor intensity and quality of mixtures of odorants presents several advantages in comparison with others like the Vectorial model, U model, UPL2 model, Additivity model, or the Euclidean additivity model (Laffort and Dravnieks, 1982; Cain et al., 1995; Teixeira et al., 2011a). This is a rough and simple model, once it does not reflect common observations in olfaction like weaker odors being able to increase or decrease the intensity of a strong odor (called odor enhancement). It is an easy method for the prediction of odor intensities of mixtures and gives some assurance that the intensity of a mixture will not be grossly larger than the odor intensity of its strongest odorant. Some authors reported that from experimental data it was also verified that this model inhibited any odorant intensity additivity effect (as expected due to its definition— Eq. (2.6)), avoiding overprediction of intensities and resulting in a quite robust model (Laffort and Dravnieks, 1982). Nevertheless, when experimental data are fitted to the SC model, reasonable results are obtained. Moreover, the SC model compares reasonably well to more complex models (Laffort and Dravnieks, 1982; Cain et al., 1995). There are several comparisons of odor perception models available in the literature: all of them show that the SC is the second or third best model in correlation with olfactory panelists' evaluations, while the best model clearly differs from experiments to experiments (Laffort and Dravnieks, 1982; Cain et al., 1995). Finally, it has the advantage of allowing the calculation of the odor intensity but also the odor character (the dominant smell) for a mixture of odorants.

Following this line of thought, if we consider a ternary fragrance mixture (e.g., a simple perfume concentrate), Eq. (2.7) reduces to Eq. (2.8):

$$OV_{mix} = \max\{OV_A, OV_B, OV_C\} \tag{2.8}$$

where subscripts represent A—top note, B—middle note, and C—base note. If a solvent is incorporated in this mixture, Eq. (2.7) can be rewritten as Eq. (2.9):

$$OV_{mix} = \max\{OV_A, OV_B, OV_C, OV_S\} \tag{2.9}$$

where subscript S represents the solvent. This particular case is for a quaternary mixture (e.g., diluted fragrance), which simulates a more realistic perfume formulation. But as the PTD® works with three components, it is necessary to recalculate the mole fractions of the fragrance components. For such a four component system, pseudoternary compositions have to be defined using compositions in a solvent-free basis, as follows:

$$x'_A = \frac{x_A}{x_A + x_B + x_C}, \; x'_B = \frac{x_B}{x_A + x_B + x_C}, \; x'_C = \frac{x_C}{x_A + x_B + x_C} \tag{2.10}$$

where x_i' indicates a pseudoternary composition for each of the three fragrant components (A, B, C) in the quaternary system (A, B, C, S).

2.1.4 Construction of the PTD®

But how do we build and interpret the PTD®? The information we want to show in the PTD® is the mapping of the perceived dominant odor of all possible ternary mixtures for some selected fragrance ingredients. From the definition of the OV (Eq. (2.1)) and considering all possible ternary mixtures within the PTD®, it is expectable that there will be some ranges of compositions (areas inside the triangle) where one fragrance component will dominate the overall odor intensity. These compositions define what is called an odor zone and where the following relationship is valid:

$$OV_i = OV_{max} > OV_{j \neq i} \tag{2.11}$$

where j is any other component in the mixture different from i. Figure 2.3 shows the PTD® for the ternary system of limonene + geraniol + galaxolide with different odor zones represented (Teixeira et al., 2009). The green region (squares) indicates the mixture compositions for which limonene has the maximum odor intensity and, thus it is the dominant odor.

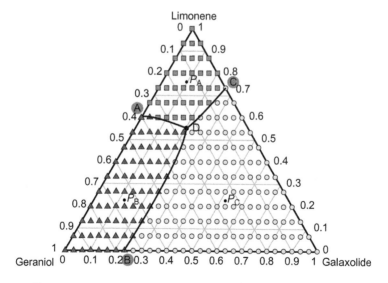

Fig. 2.3. PTD® for the ternary system limonene + geraniol + galaxolide considering nonidealities in the liquid phase. The colors distinguish the regions (compositions) with different dominant odors. These regions are separated by lines, the so called PBL. Adapted with permission from Teixeira et al. (2009). © 2009, John Wiley & Sons.

Similarly, the pink (triangles) and yellow (circles) regions denote the compositions for which geraniol and galaxolide, respectively, are the dominant odors. As it can be seen in the diagram, three different odor zones can be distinguished where the aforementioned relationship can be summarized as:

P_A—The dominant odor is limonene (top note): $OV_{max} = OV_A$.
P_B—The dominant odor is geraniol (middle note): $OV_{max} = OV_B$.
P_C—The dominant odor is galaxolide (base note): $OV_{max} = OV_C$.

Also shown in Fig. 2.3 are three important lines $(\overline{AD}, \overline{BD}, \overline{CD})$ that represent the limits of the odor zones and are called perfumery binary lines (PBL). In such lines, the maximum odor intensity is shared by two components and their calculation is critical for mapping the PTD$^{\circledR}$. Another singular point in the PTD$^{\circledR}$ is point D: a single ternary composition where three PBLs intersect. It is called the perfumery ternary point (PTP) and for that specific composition, the three components share the maximum odor intensity. If a solvent or other fragrance ingredient is introduced in the mixture, there may be compositions where three fragrance components will share the same OV, though this will be lower than the one for the fourth component. This is simply called a ternary point and is of value when the same olfactive target is obtained for all the fragrance ingredients but the solvent odor intensity prevails (Mata et al. 2005c).

The simplest approach for the construction of the PTD$^{\circledR}$ is the application of Eqs (2.5) and (2.7) to a given set of compositions. That would provide the dominant odor for those compositions. However, for practical reasons it is better to calculate the limits of the odor zones that is the PBLs and PTPs mentioned above.

In order to obtain the compositions of the PBL between two given components i and j, it is necessary to solve a system of two equations with some constraints:

$$OV_i = OV_j \tag{2.12}$$

$$\sum_{i=1}^{3} x_i = 1 \tag{2.13}$$

with the following constraints:

$$\{OV_i, OV_j\} > OV_k \quad \forall i, \; j \neq k \tag{2.14}$$

Table 2.1. Conditions for the Determination of PBL		
Type I $i+j$	**Type II** $i+k$	**Type III** $k+j$
$OV_j = OV_i$ $x_i + x_j + x_k = 1$ $\{OV_i = OV_j\} = OV_{max}$ $0 < x < 1$	$OV_i = OV_k$ $x_i + x_j + x_k = 1$ $\{OV_i = OV_k\} = OV_{max}$ $0 < x < 1$	$OV_k = OV_j$ $x_i + x_j + x_k = 1$ $\{OV_k = OV_j\} = OV_{max}$ $0 < x < 1$
Adapted with permission from Teixeira (2011). © 2011, John Wiley & Sons.		

where subscripts i and j stand for the two components of the PBL. The procedure must be repeated for all possible pairs of components. For this typical ternary system with three fragrance notes (top, middle, and base notes), there are three possible PBLs.

Table 2.1 presents the conditions for their determination in the ternary system. In general, for a perfumery system with N components, there will be C_2^N possible combinations of PBLs. As said before, there are also some composition points where two fragrance components share the same OV but that is lower than the maximum OV of the mixture, i.e., $OV_i = OV_j < OV_{max}$ (for $i, j = A, B, C, S$) (Mata et al., 2005c; Mata and Rodrigues, 2006).

The calculation of the PTP is the extension to three components of the previous analysis. Considering a PTP between components i, j, and k, it is determined by solving a system of equations:

$$OV_i = OV_j = OV_k \qquad (2.15)$$

$$\sum_{i=1}^{3} x_i = 1 \qquad (2.16)$$

In general, for a perfumery system with N components, there will be C_3^N possible combinations of PTPs. In the ternary system evaluated here, there will be only one possible type of PTP. It defines the specific composition of the mixture where the top, middle, and base notes share the maximum OV (OV_{max}) of the mixture. As the reader may understand, the odor of such a mixture is different from the individual components alone, because there will be a mixture of perceived odorants with indistinguishable intensities. More details on these calculations have been given in previous articles of the group (Teixeira et al., 2009, 2010).

2.1.5 The Relevance of the Activity Coefficient (γ)

Before getting into details with the application of perfumery systems using the PTD®, we will make a brief parenthesis here to discuss one important topic that must be addressed. From inspection of Eq. (2.5), one of the parameters presented there is the activity coefficient (γ_i) which can be estimated using the well-known UNIFAC method (among others), as aforementioned. The calculation is straightforward, but tedious, especially if you have endless possible mixtures. So, how important is it for the estimation of odor intensity? From a theoretical point of view, it is necessary, considering the variety of functional groups present in fragrance molecules (not to mention the use of polar solvents). But may it be questioned from a perfumery (and qualitative) point of view?

The classification of single fragrance chemicals can be done in terms of their volatilities (top, middle, base notes). However, when diluted in a mixture, the behavior of a fragrance, and consequently its OV, is also a function of its interactions with the other chemical species present in the solution. These interactions are due to differences in molecular size and in a great extent to energetic interactions (London forces, interactions produced by aromatic rings, dipoles, hydrogen bonding, among others) (Poling et al., 2004). As mentioned before, the affinity of a molecule for its surrounding medium can be measured using the concept of the activity coefficient, γ. Behan and Perring (1987), showed the large differences that can be observed in the activity coefficient of benzyl acetate (i), heptan-2-ol (j), and limonene (k) when they are diluted in different solvents like water (high polarity), an aqueous surfactant, or diethyl phthalate (DEP–a moderately polar solvent). In the case of a DEP solution, all fragrance solutes presented relatively low values for the activity coefficient ($\gamma_i = 0.3$; $\gamma_j = 2.2$; $\gamma_k = 2.3$), while moderately high values were observed in the aqueous surfactant ($\gamma_i = 75$; $\gamma_j = 57$; $\gamma_k = 732$), and very high values in water ($\gamma_i = 1750$; $\gamma_j = 2770$; $\gamma_k > 70{,}000$). The exceptionally high value of limonene activity coefficient in water means that its nonpolar molecules do not tend to participate in any interactions with water. Consequently, limonene molecules will tend to leave the liquid mixture into the gas phase, thus increasing limonene headspace concentration and its perceived odor. This shows why molecular interactions and activity coefficients are very important topics inside perfumery, and more specifically in perfume formulation and odor performance, once the quantity of alcohol, water, or other solvent/cosmetic base can largely influence the headspace concentration above the liquid.

However, it should be pointed out that although the vapor composition of a fragrance mixture may significantly deviate from ideality (which means neglecting activity coefficients, $\gamma = 1$), in terms of the perceived odor it may not change much. At first sight, this may look controversial, but let us explain why. If we look at Eq. (2.5) for the calculation of OV, we see that the composition in the liquid phase (x_i) appears in the numerator and is multiplied by the activity coefficient (γ_i). It is often observed that a high dilution factor of a solute within a mixture $(x_i \rightarrow 0)$, causes a significant deviation from ideality $(\gamma_i$ largely deviates from unity). Conversely, when a component is present in high concentration $(x_i \rightarrow 1)$, the system tends to behave like an ideal one, so the activity coefficient approximates to unity $(\gamma_i \rightarrow 1)$. A comparison between the variation of the odor intensity (in terms of OV) for the ideal and nonideal solution (using the UNIFAC prediction) of a binary mixture of limonene + geraniol is shown in Fig. 2.4.

It is clearly seen that the behavior of this binary system deviates from ideality, and so does the predicted OV for each of the fragrance ingredients. However, it is also seen that, from an odor quality perspective, the point where the dominant odor changes from limonene to geraniol (or the opposite) occurs for a similar mixture composition $(x_{\mathrm{Ger}} \approx 0.4)$. In this way, a prediction of the dominant odor as well as

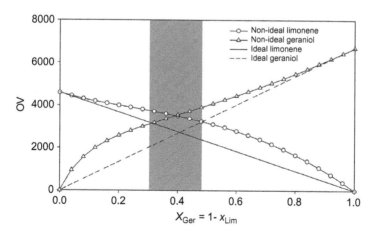

Fig. 2.4. Variation of OV for binary mixtures of limonene and geraniol. Comparison between ideal model (I) and nonideal model (NI).

the perceived odor for the mixture near the switching point (grey band) is very similar considering both the ideal and the nonideal approaches. However, nonidealities may have greater implications on diluted mixtures of fragrances in different types of solvents. For that purpose, a comparison is made in Fig. 2.5 between ideal and nonideal mixtures of the type (limonene + geraniol + galaxolide + ethanol) with and without solvent.

If the behavior of the ternary mixtures in Fig. 2.5 (top diagrams) is similar when considering the ideal and nonideal approach, the same does not happen for diluted mixtures (Fig. 2.5, bottom). When ethanol is incorporated in the solution, it is seen that the ideal and nonideal predictions are clearly different. According to the nonideal prediction, limonene is much more pushed out from the mixture, and so it is more strongly perceived (thus its odor zone is larger). Furthermore, when the ideal approach is considered there is a small odor zone for ethanol in the middle of the diagram which is not detected when nonidealities are considered. Such behavior has been seen for other perfumery systems as well, showing that for multicomponent mixtures, especially those diluted, the effect of nonidealities is of relevance.

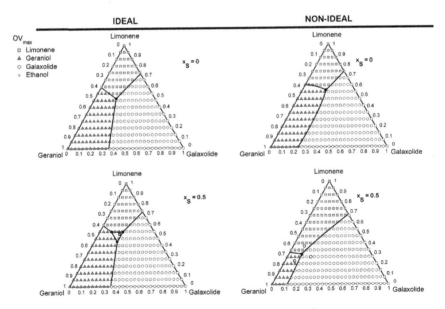

Fig. 2.5. *Effect of nonidealities in the perceived odor character mapped in the PTD® for ternary (top) and quaternary mixtures ($x_S = 0.5$, bottom). Left diagrams are for ideal and right diagrams are for nonideal mixtures. PTPs (PTP, •) and Ternary Points (TP, o) are represented.* Adapted with permission from Teixeira et al. (2009). © 2009, John Wiley & Sons.

A different thing is the quantitative prediction of the odor intensity of mixtures. Should the inclusion of nonidealities in our modeling be of significant importance in the prediction of odor intensities for perfume mixtures? In order to assess that effect, we have compared the distribution of OVs for all possible ternary and quaternary (diluted) mixtures of the previous perfumery system. The predicted planes of dominant OVs for such mixtures are presented in Fig. 2.6 for the ideal (left) and nonideal (right) cases.

It is possible to see that for ternary mixtures (bottom PTD of the triangular prisms), there are only slight differences in the prediction of the odor intensities using both the ideal and the nonideal approaches. However, when ethanol is included in the mixture, larger differences arise: if we consider an ideal mixture, the ethanol odor intensity will be more strongly perceived than fragrance ingredients as we increase its concentration in the liquid. Thus, for the ideal case, introduction of ethanol lowers the odor intensity of fragrances more significantly than when nonidealities are considered (as can also be seen in Fig. 2.5). This issue is certainly very relevant from the perfumery point of view, since when designing a fragrance it is not desired to perceive a strong alcoholic odor.

As a conclusion, there will be mixtures where ideality represents fairly well its thermodynamic behavior and, consequently, its released odor, though for many others that is unlikely to happen. As we will see later in this book, some multifunctional fragrance ingredients will introduce significant changes in the perceived odor of multicomponent

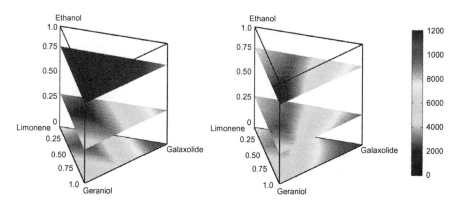

Fig. 2.6. Effect of nonidealities in the perceived odor intensity (OV) for ternary mixtures and quaternary mixtures (ideal—left, nonideal—right). Adapted with permission from Teixeira (2011). © 2011, M.A. Teixeira.

mixtures. In this way, accounting for the effect of intermolecular interactions is important and should be considered, especially when multicomponent fragrance ingredients are presented and diluted in suitable perfumery solvents like ethanol or water.

2.2 APPLICATION OF THE PTD® METHODOLOGY: EFFECT OF BASE NOTES

Here, we will show the application of the PTD® methodology for different ternary and pseudoquaternary perfumery systems, containing top, middle, and base notes with and without considering the presence of solvents. A comparison on the effect of different base notes in the perceived odor will be used to exemplify its application. The selected odorants are presented in Table 2.2 and Fig. 2.7, together with their corresponding chemical structures and physical and sensorial properties.

Figures 2.8−2.10 present the PTDs for the three different perfumery systems: limonene + geraniol + base note + ethanol. The base notes used were tonalide, vanillin, and galaxolide. The figures show the obtained PTDs for concentrated perfume mixtures (ethanol free) and for quaternary mixtures with some selected compositions of ethanol (30, 50, and 70 mol%). The PTDs for each system and ethanol composition show the different odor zones together with the PBL.

The PTD® methodology together with the prediction of the activity coefficients (γ_i) using the UNIFAC method was implemented in the

Table 2.2. Properties of the Odorant Components						
	Name	Molecular Formula	M_i (g/mol)	P_i^{sat} (Pa)	ODT_i (g/m³)	$\dfrac{P_i^{sat} \cdot M_i}{ODT_i \cdot RT}$
A	Limonene[a]	$C_{10}H_{16}$	136.2	20.5×10^1	2.45×10^{-3}	4.60×10^3
	α-Pinene[a]	$C_{10}H_{16}$	136.2	5.13×10^2	5.44×10^{-4}	5.18×10^4
B	Geraniol[a]	$C_{10}H_{18}O$	154.3	26.7×10^{-1}	2.48×10^{-5}	6.70×10^3
	Linalool[a]	$C_{10}H_{18}O$	154.3	2.21×10^1	1.26×10^{-5}	1.09×10^5
C	Vanillin[a]	$C_8H_8O_3$	152.2	16.0×10^{-3}	1.87×10^{-7}	5.25×10^3
	Tonalide[a]	$C_{18}H_{26}O$	258.4	67.0×10^{-6}	1.82×10^{-5}	3.84×10^{-1}
	Galaxolide	$C_{18}H_{26}O$	258.4	72.7×10^{-3b}	6.30×10^{-7c}	1.20×10^4
S	Ethanol[a]	C_2H_6O	46.0	72.7×10^2	5.53×10^{-2}	2.44×10^3

[a] From Calkin and Jellinek (1994).
[b] From Balk and Ford (1999).
[c] From Fráter et al. (1999).
Adapted with permission from Teixeira et al. (2009). © 2009, John Wiley & Sons.

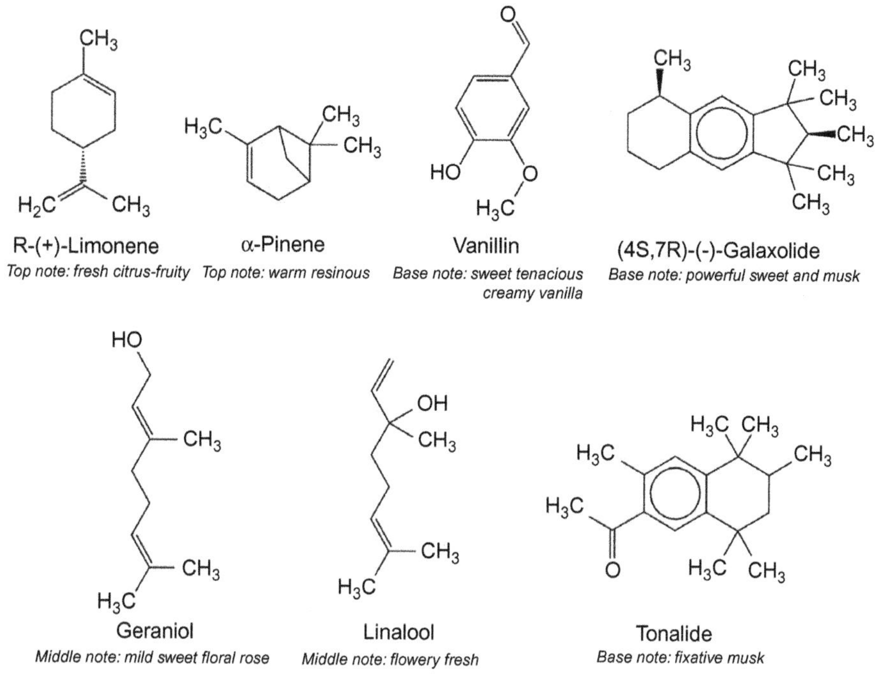
CH3
H2C
CH3
R-(+)-Limonene
Top note: fresh citrus-fruity
H3C
CH3
CH3
α-Pinene
Top note: warm resinous
O
HO
O
H3C
Vanillin
Base note: sweet tenacious creamy vanilla
CH3
CH3
CH3
CH3
CH3
CH3
(4S,7R)-(-)-Galaxolide
Base note: powerful sweet and musk
HO
CH3
CH3
H3C
Geraniol
Middle note: mild sweet floral rose
OH
CH3
CH3
H3C
Linalool
Middle note: flowery fresh
H3C CH3
H3C
CH3
H3C
O
H3C CH3
Tonalide
Base note: fixative musk

Fig. 2.7. Chemical structures of the fragrant components.

MATLAB software, incorporating the routines for resolution of all the simulations. Moreover, due to the application of the UNIFAC method, the system of equations turns to be nonlinear and so has to be computed numerically using iterative methods. This was performed within the optimization toolbox package from MATLAB, adapted to the nature of the problem studied here (Chapman, 2000; Chapra and Canale, 2002). The nonlinear system of algebraic equations was solved using the numerical method of Levenberg–Marquardt with line search. The choice of this method relied on its robustness and precision, although it may have occasionally poorer efficiency than other methods (e.g., higher number of function evaluations than the Gauss–Newton method, when the residual is zero at the solution). The line search algorithm uses a combined quadratic and cubic polynomial interpolation and extrapolation methods (Nocedal and Wright, 1999; MathWorks, 2002).

Inspection of Figs 2.8–2.10 and focusing in the ternary systems only ($x_S = 0$), the first conclusion is withdrawn: changing only one

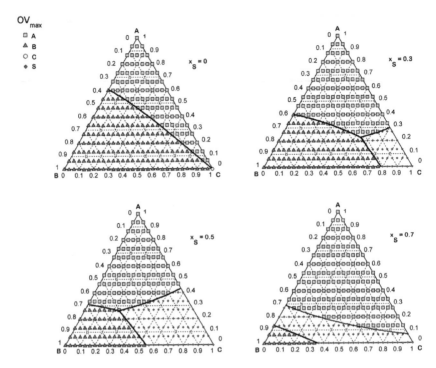

Fig. 2.8. PTDs for different initial mole fractions of solvent in the quaternary system of limonene (A) + geraniol (B) + tonalide (C) + ethanol (S). Adapted with permission from Teixeira et al. (2009). © 2009, John Wiley & Sons.

component in the mixture (here, the base note) the perceived odor may drastically change. In the three cases presented here, the predicted odor zones show completely different shapes for the dominant odors. Moreover, considering each perfumery system alone, the PTDs can be seen in two perspectives:

1. *From top to bottom*: Increasing the ethanol mole fraction in the liquid mixture, it is possible to determine its maximum concentration so that the OV_{max} corresponds to the desired fragrant component but not to ethanol. An example happens when sniffing a perfume bottle: it is expected that only the fragrance ingredients are perceived, but never the strong alcoholic-ethereal-medical-like smell of ethanol.

2. *From bottom to top*: Decreasing the ethanol mole fraction in solution, we are simulating its evaporation, which usually takes some few seconds (up to few minutes) after application. It is like when we spray a perfume on a paper blotter, smell it, fan it, and smell it again.

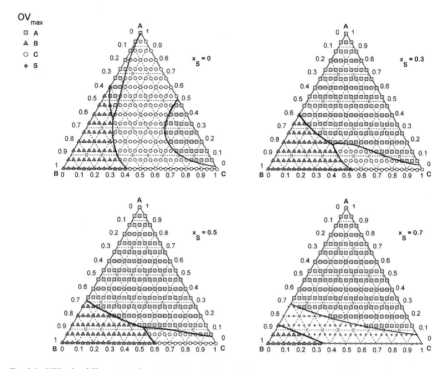

Fig. 2.9. PTDs for different initial mole fractions of solvent in the quaternary system of limonene (A) + geraniol (B) + vanillin (C) + ethanol (S). Adapted with permission from Teixeira et al. (2009). © 2009, John Wiley & Sons.

The first perfumery system (Fig. 2.8) uses tonalide as base note. Tonalide is a typical sweet-musk-warm fragrance, commonly used in perfumery as a fixative although its application is currently more restricted. A fixative is a chemical ingredient (usually of low volatility) that can change the rate of evaporation of both the top and the middle notes and improve stability, thus allowing the perfume to last longer (Calkin and Jellinek, 1994; Teixeira et al., 2009; Ohloff et al., 2012). In this perfumery system, the predicted PTD® shows that tonalide is only perceived when pure (corner C), while limonene and geraniol divide the diagram in two odor zones (when $x_S = 0$). Moreover, the introduction of ethanol in this mixture brings its corresponding odor zone that expands from corner C: this means also that tonalide is not a potent odorant and is fixating the remaining fragrances in the mixture.

On the perfumery system of Fig. 2.9, the base note is changed to vanillin, highlighting a curious and unexpected behavior. In what concerns the corresponding ternary mixture, not only the odor zone for

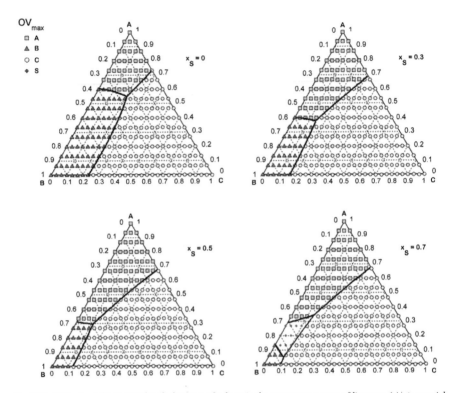

Fig. 2.10. PTDs for different initial mole fractions of solvent in the quaternary system of limonene (A) + geraniol (B) + galaxolide (C) + ethanol (S). Adapted with permission from Teixeira et al. (2009). © 2009, John Wiley & Sons.

vanillin exhibits an unusual shape but there are also two different odor zones for limonene (one at high and another at low concentrations). The explanation behind the fact that both limonene and vanillin present a strong odor, even when in very low concentrations in the liquid phase, is given by Thermodynamics. This effect is explained by the relative low polarity of limonene, which is highly "pushed out" of the polar solution at low concentrations, thus increasing its perceived odor intensity in air. A similar phenomenon is predicted for the odor intensity of vanillin (as seen in Fig. 2.9), thus showing a good agreement with experimental observations, since it is known to be a powerful base note (TGSC, 2012).

Finally, the perfumery system presented in Fig. 2.10 shows that according to our predictions, although galaxolide is a typical base note, it is strongly perceived near the gas—liquid interface for a large range of compositions in these mixtures. A closer look at the physico-chemical properties of this component (see Table 2.2) shows that despite having a low saturation pressure, the ODT is also very low, which leverages the potency of its odor. This is why perfumers usually use galaxolide highly diluted in the perfume formulation (it often comes diluted in DEP for product stability as well).

In sum, the PTD® allows predicting and evaluating the different properties of the base notes studied here, showing that the effect into the odor character increases in the order of: tonalide ≪ vanillin < galaxolide (Figs 2.8–2.10, respectively).

Another feature of the PTD® is that it is also able to show the odor intensity of a fragrance ingredient by plotting the OVs in the ternary diagram. Like this, it is possible to map the contours of odor intensity for each component (like isolines of OV) as presented for the system limonene (A) + geraniol (B) + vanillin (C) in Fig. 2.11. Regarding that example, it is important to make some considerations: while for gera-niol (middle note), the maximum odor intensity is reached when it is presented at high concentrations in the mixture (the expected behavior), the same is not seen for limonene (top note) or vanillin (base note). In fact, the behavior of these two fragrance ingredients is quite curious as previously discussed. For ternary mixtures the predicted odor intensity for limonene reaches its maximum at low concentrations, with high con-centrations of vanillin. The opposite is found for vanillin. Once again, the explanation behind this behavior is in the activity coefficient, and the ability of a polar solution to push out the nonpolar fragrances, and vice versa.

However, addition of ethanol (in a mole fraction of $x_S = 0.30$, 0.50, 0.70), not only changes significantly the shape of the PTD® (as shown

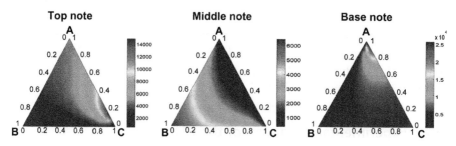

Fig. 2.11. PTDs showing the odor intensity map for each fragrance ingredient in the ternary mixture of limonene (top note) + geraniol (middle note) + vanillin (base note).

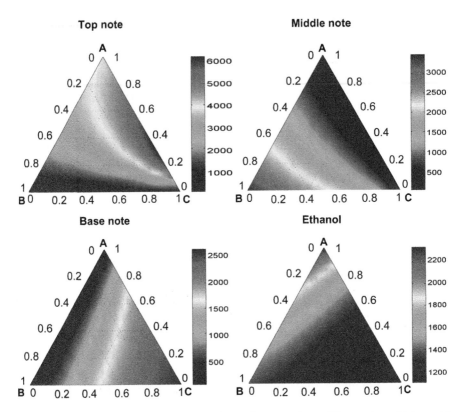

Fig. 2.12. Odor intensity map of each fragrance ingredient in the ternary mixture of limonene (top note) + geraniol (middle note) + vanillin (base note) + ethanol ($x_S = 0.50$, solvent).

in Fig. 2.9) but also the odor intensity of each component. This is presented in Fig. 2.12 for $x_S = 0.50$. In this case, the OVs decrease significantly for all fragrance compounds mainly due to dilution effects. The incorporation of ethanol increases the polarity of the liquid mixture. Consequently, vanillin is more retained in solution at low concentrations, which decreases its vapor concentration and so, its odor intensity. On the contrary, the OVs for limonene still continue to be significantly high when present in low concentrations: This happens because limonene is nonpolar, and the addition of ethanol keeps the liquid solution highly polar. Thus, limonene is still pushed out of the mixture at low concentrations.

2.2.1 Experimental Validation of the PTD®

As a proof of concept, it is important at this stage to evaluate the experimental validation of the predictive PTD® model. This can be

Table 2.3. Composition of Perfume Test Mixtures (Compositions Are Presented in Molar Fractions)

		Molar Liquid Composition		
	Component	P1	P2	P3
A	Limonene	0.129	0.289	0.096
B	Geraniol	0.129	0.145	0.339
C	Vanillin	0.129	0.048	0.048
S	Ethanol	0.613	0.518	0.517
Adapted with permission from Gomes et al. (2008). © 2008, John Wiley & Sons.				

done by measuring OVs for some mixtures, based on experimental ODT values and headspace concentrations (C^g). One of the previous systems addressed was used for the validation: limonene (A) + geraniol (B) + vanillin (C) + ethanol (S). Here, we will show it using three simple quaternary perfume mixtures by performing a comparison between the experimental OVs with those predicted from our simulations for the ideal and nonideal approaches. The compositions of these test mixtures are given in Table 2.3.

The experimental OVs are calculated using Eq. (2.1), with the corresponding experimental ODTs and headspace concentrations (in equilibrium with the liquid phase) for each component. For that purpose, panelists were used for the determination of ODTs (by dynamic olfactometry), while gas concentrations were obtained by static headspace GC analyses of the vapor phase above the liquid mixture at 25°C in equilibrium conditions (Gomes et al., 2008). From the latter experiments, it is also possible to calculate an approximated saturated vapor pressure for single components using the ideal gas law:

$$P_i^{sat} = \frac{C_i^g}{M_i} RT \tag{2.17}$$

These experimental saturated vapor pressures together with the experimental ODTs were used for the validation of the PTD® model. Figure 2.13 shows a comparison between experimental and calculated OVs (ideal and nonideal) obtained for the three test mixtures (P1, P2, and P3). It is seen that limonene presents the maximum OV for all three mixtures considering the experimental OVs. This was confirmed by olfactory evaluations performed for the three different test mixtures with limonene being the dominant odor. Additionally, it is to note that

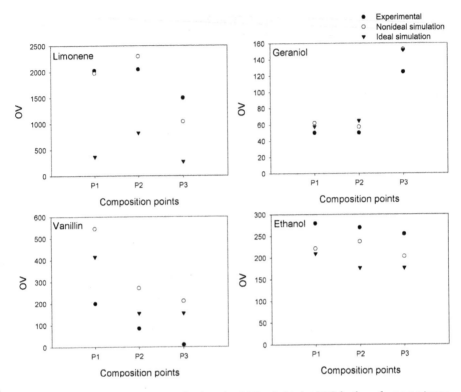

Fig. 2.13. Comparison between experimental and simulated OVs calculated at 25°C for the perfume test mixtures with limonene, geraniol, vanillin, and ethanol (composition points P1, P2, and P3). The simulation was performed for both ideal and nonideal cases.

the OVs for limonene, geraniol, and ethanol (considering nonidealities, open circles) were reasonably well predicted. In the case of geraniol, the activity coefficients (and the OVs) are often slightly over predicted, while the opposite happens for ethanol. For the case of vanillin, a molecule with multifunctional groups and very polar, there are large deviations, especially for mixture P1 (high concentration of vanillin). While the experimental OVs for the previous components differed from those calculated in a range of 2.5–30%, in the case of vanillin, the difference was significantly higher. Moreover, it is apparent from these results that a special interaction between the limonene–vanillin pair happens. The evaporation of vanillin seems to be somehow enhanced in the presence of limonene at relatively high concentrations (P1 and P2). As we have seen before, vanillin is a perfumery material difficult to work with, although it is one of the most important ones and makes part of countless perfumes (Calkin and Jellinek, 1994). In a perfume with a concentration as low as 0.5% of vanillin, it can overlay the odor of the remaining components. Therefore, it is usually the last ingredient to be added and the most effective concentration level is often established by trials (Gomes et al., 2008). Finally, it should be noted, once more, the importance of considering nonidealities in the liquid phase: the predictions using the ideal model (neglecting activity coefficients) are considerably worse in respect to the experimental values. We have previously addressed this issue in Section 2.1.2. The relevance of the activity coefficient (γ). Here, we validate our opinion using olfactory experimental data as well.

In sum, it should be said that although a perfume is formulated in the liquid phase, it is the headspace concentration that will define its smell. The OVs were reasonably well predicted by the PTD® methodology except for vanillin, but the dominant odorant was well predicted for the three test mixtures. Thus, this methodology shows to be useful in predicting odor intensities and the dominant character of perfume mixtures: consequently, it may be helpful in the preformulation stages of simple fragrances like those incorporated in detergents, soaps, or candles.

2.3 THE PERFUMERY QUATERNARY–QUINARY DIAGRAM (PQ2D®)

The previous PTD® methodology can be used to represent the odor character of ternary fragrance mixtures. Additionally, it can also be used for quaternary mixtures if the fourth component has a fixed composition (and so, the remaining components are represented as in a solvent-free basis). Despite it may serve to provide some insights on the behavior of quaternary mixtures, it is also true that this type of projection has its inherent limitations. It would be necessary a series of PTDs with different (fixed) compositions of solvent to show the whole behavior of a quaternary mixture (e.g., as in a triangular prism). In order to overcome that issue, the previous PTD® methodology was extended to a new tool called Perfumery Quaternary-Quinary Diagram (PQ2D®). This PQ2D® maps the odor space of quaternary to quinary mixtures. It follows its predecessor in much part of the methodology but because it uses three-dimensional representations, it allows showing the perceived odor of quaternary mixtures. For that purpose, a regular tetrahedron is used where each vertex represents a single component of the mixture: the triangular base of the tetrahedron can be seen as the previous PTD® (as well as its faces) while at the top there will be the fourth component. Like in the ternary diagram, it is imperative that for each point inside the tetrahedron, the sum of the mole fractions equals to unity. In fact, the regular tetrahedron possesses an important property: if we consider a point inside the tetrahedron, the summation of the distances from that point to all four faces (e.g., the sum of the heights of that point in respect to each face) is a constant value, k, which equals the total height of the tetrahedron (Wei, 1983):

$$\sum_i h_i = k \qquad (2.18)$$

where h_i is the distance from a point inside the tetrahedron to the face i. This is analogous to the equilateral triangle, used in the PTDs. In this way, if we build a tetrahedron of unit length edges (so the edges can be used as axis to denote fractions of compositions), we can use it to represent the composition (in mole, mass, or volume fraction) of any quaternary system, given that:

$$\sum_i x_i = 1 \propto \sum_i h_i \qquad (2.19)$$

where x_i stands for the composition of component i (in mole, mass, or volume fraction). Following this line of thought, each edge of the tetrahedron represents a binary subsystem (composed by the two pure components at both ends of the edge), each face represents a ternary subsystem (without the component placed in the opposite vertex) and, finally, the points inside the tetrahedron represent any possible mixture of the quaternary system. Moreover, all points in a plane parallel to a given face are at the same distance of that face and, thus, have the same composition for the component in the opposite vertex to that face. It should be noted that the regular tetrahedron is commonly used in phase equilibria thermodynamics to represent the behavior of quaternary systems (Wei, 1983).

Once more, in analogy with the ternary diagrams, the PQ2D® also needs a transformation of coordinates from the system of tetrahedric coordinates (quaternary compositions, in mole, mass, or volume fraction) to the three-dimensional, cartesian space. (Walas, 1985; Teixeira et al., 2009).

Moreover, the PQ2D® may use the concept of OV as defined in Eq. (2.1) as well as the SC model, thus mapping the dominant odor within the tetrahedric diagram. Analogously to the PTD® where we had odor zones, now in the PQ2D® there will be ranges of compositions called odor volumes where one component dominates the overall odor. For a quaternary mixture, the compositions confined to these odor volumes must follow the relationship:

$$OV_i = OV_{max}, \quad i = A, B, C, S \tag{2.20}$$

In this way, an odor volume is limited by surfaces where the OV_{max} is shared by two components and delimited by lines where the OV_{max} is shared by three components. So, a new concept of perfumery binary surfaces (PBS) and perfumery ternary lines (PTL) is introduced with the PQ2D®. The PBS defines the region of the quaternary compositions where:

$$OV_i = OV_j = OV_{max}, \quad i,j = A, B, C, S \tag{2.21}$$

with i, j as fragrant components (being $i \neq j$). The PTL are the set of compositions where the following relationship remains valid:

$$OV_{max} = OV_i = OV_j = OV_k, \quad i,j,k = A, B, C, S \tag{2.22}$$

with i, j, k as fragrant components and (being $i \neq j \neq k$). The PTL results from the intersection of two different PBS. This means that a PTL crosses the quaternary diagram in the thin region where three fragrant components have the same maximum OV (OV_{max}). Moreover, in a quaternary mixture, it is possible to have one or more compositions where all the components share the maximum OV. That is defined as a perfumery quaternary point (PQP) between four different fragrant components i, j, k, and l when they share the maximum OV:

$$OV_i = OV_j = OV_k = OV_l = OV_{max}, \quad i,j,k,l = A, B, C, S \tag{2.23}$$

where i, j, k, and l are fragrant components (being $i \neq j \neq k \neq l$). Similarly, the composition of the PQP is also set by the intersection of three different PTLs. For a quaternary perfumery system, there will be $C_2^4 = 6$ different possible combinations of PBS, $C_3^4 = 4$ different possible combinations of PTLs, and there is only one type of PQP. The conditions for the determination of PBS, PTL, and PQP considering a

Table 2.4. Conditions of the Nonlinear System for the PBS, PTL, and PQP		
PBS $i+j$	PTL $i+j+k$	PQP $i+j+k+l$
$OV_i = OV_j$	$OV_i = OV_j = OV_k$	$OV_i = OV_j = OV_k = OV_l$
$x_i + x_j + x_k + x_l = 1$	$x_i + x_j + x_k + x_l = 1$	$x_i + x_j + x_k + x_l = 1$
$\{OV_i = OV_j\} = OV_{max}$	$\{OV_i = OV_j = OV_k\} = OV_{max}$	$\{OV_i = OV_j = OV_k = OV_l\} = OV_{max}$
$0 < x < 1$	$0 < x < 1$	$0 < x < 1$
Adapted with permission from Teixeira et al. (2009). © 2009, John Wiley & Sons.		

quaternary perfume system are shown in Table 2.4. The determination of these PBS, PTL, and PQP, as previously happened with the PTD® methodology, includes the numerical calculation of a large number of possible composition points.

2.3.1 Application of the PQ2D® Methodology

Similarly to what was done with the PTD®, the application of the PQ2D® methodology will be shown using the previous three different quaternary systems: limonene (A) + geraniol (B) + base note (C) + ethanol (S), with the base notes of tonalide, vanillin, and galaxolide. Before the tetrahedrons are discussed, their faces will be presented and evaluated. The faces of the tetrahedron represent the ternary subsystems that compose the quaternary system. In this way, for a perfumery system of the type (A + B + C + S), the faces will be the four ternary subsystems: (A + B + C), (A + B + S), (A + C + S), and (B + C + S). This allows an easier perspective for visual interpretation than with the three-dimensional projections. These faces of the tetrahedron are shown in Fig. 2.14 for the three different quaternary systems.

Each diagram is divided in four triangles (PTD®) which represents a single face of the tetrahedron, i.e., each of the four different ternary subsystems of the quaternary mixture. Moreover, from these projections the tetrahedron can be constructed from each diagram just by "pushing up" the three outer vertices (S), so that they join together in the space above the central triangle (base of the tetrahedron). It is important to mention that Fig. 2.14 only shows the limits (faces) of the tetrahedron, thus no information about the behavior of the quaternary system is provided. In order to visualize that, it is necessary to plot the interior composition points of the quaternary mixtures. This is shown in Figs 2.15–2.17, where the odor volume for each component is represented separately, so that its shape can be clearly visualized.

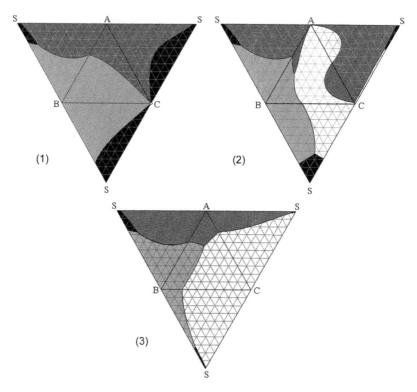

Fig. 2.14. Ternary subsystems for the three perfumery systems: (A) top note (dark gray, limonene); (B) middle note (light grey, geraniol); (C) base note (white, 1: tonalide, 2: vanillin, 3: galaxolide); (S) solvent (black, ethanol). The figures are the projection of the tetrahedron faces. Adapted with permission from Teixeira et al. (2009). © 2009, John Wiley & Sons.

The complementarity between Fig. 2.14 and the corresponding tetrahedrons (Figs 2.15–2.17) is essential for an easy visualization of the odor space of the perfumery systems. From these PQ2Ds, it is possible to understand the whole behavior of the quaternary perfumery systems, something which was not possible using only the previous PTDs (compare Figs 2.8 and 2.15; Figs 2.9 and 2.16; and Figs 2.10 and 2.17).

Indeed, the ternary diagrams represent horizontal cuts of these PQ2Ds, at different heights from the bottom of the tetrahedric diagram (i.e., for different ethanol compositions). In this way, the odor volumes obtained for these PQ2Ds confirm what could be hypothesized from inspection of the PTD® diagrams in Figs 2.8–2.10, or from the ternary subsystems in Fig. 2.14: tonalide is only the dominant note when nearly pure, since its odor volume is restricted to its corresponding vertex. The odor volume of vanillin confirms its unusual shape,

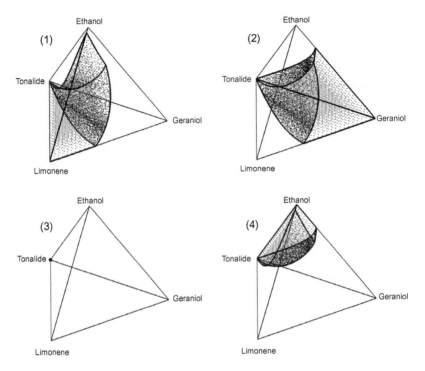

Fig. 2.15. Perfumery fragrance volumes for the quaternary system limonene + geraniol + tonalide + ethanol: (1) limonene, (2) geraniol, (3) tonalide, (4) ethanol.[1] Adapted with permission from Teixeira et al. (2009). © 2009, John Wiley & Sons.

being a dominant note both in its high and low concentration range. Furthermore, galaxolide presents the largest of the odor volumes, dominating most of the tetrahedric diagram (and thus, most of the composition spectrum).

An important result to extract from these PQ2Ds is that the potency of the base notes can be seen by the size of the corresponding odor volume. As previously seen for the PTDs, it increases in the order of tonalide $<<<$ vanillin $<$ galaxolide. This phenomenon is related with the different blooming of the base notes and deserves further attention. A relatively common mistake found in the literature states that the potency of an odorant or its OV can be simply approximated by the saturated vapor pressure of the odorant. Nevertheless, that is a gross approximation that may lead to erroneous conclusions. Recalling the

[1]All these PQ2D® can be seen in animated movies for their whole odor distribution at www.lsre.fe.up.pt.

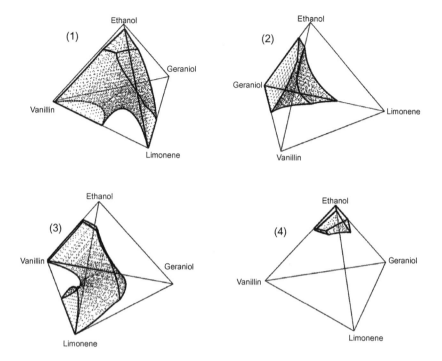

Fig. 2.16. Perfumery fragrance volumes for the quaternary system limonene + geraniol + vanillin + ethanol: (1) limonene, (2) geraniol, (3) vanillin, (4) ethanol. Adapted with permission from Teixeira et al. (2009). © 2009, John Wiley & Sons.

definition of the OV as expressed by Eq. (2.5), it is possible to rewrite it as:

$$OV_i = \gamma_i \cdot x_i \cdot \left(\frac{P_i^{sat} \cdot M_i}{ODT_i \cdot RT} \right) = \gamma_i \cdot x_i \cdot K_i^{odor} \qquad (2.24)$$

where K_i^{odor} represents the ratio of properties inside parenthesis. Among the three factors from Eq. (2.24), it is seen that the composition (x_i) varies from 0 to 1, while the activity coefficient (γ_i) can vary in most cases about one or two orders of magnitude. On the other hand, the K_i^{odor} is a constant value for each fragrance chemical. It is a function of the experimental temperature (T) and some pure component properties (saturated vapor pressure, molecular weight, and ODT). Table 2.2 presented the values for this constant (K_i^{odor}) calculated for each odorant used in the perfumery systems studied here. It is clearly seen that it increases in the same way as the size of the odor volumes for the corresponding base notes. Moreover, as discussed

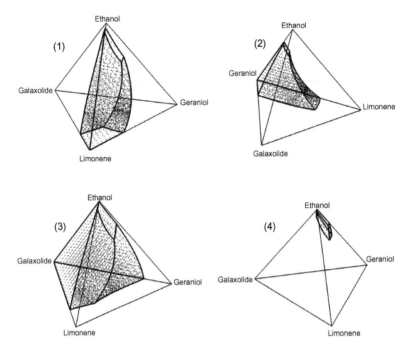

Fig. 2.17. Perfumery fragrance volumes for the quaternary system limonene + geraniol + galaxolide + ethanol: (1) limonene, (2) geraniol, (3) galaxolide, (4) ethanol. Adapted with permission from Teixeira et al. (2009). © 2009, John Wiley & Sons.

before, the product of the composition by the activity coefficient tends to have a lower impact on the OV than the K_i^{odor}. That is especially significant from tonalide to vanillin, which may explain why the first cannot be strongly perceived when mixed with other fragrant components. It is to highlight also that tonalide has the lowest saturated vapor pressure and the highest odor threshold, resulting in a K_i^{odor} that is several orders of magnitude lower than all others. K_i^{odor} for galaxolide, for its part, explains why the predictions point to its strong perception for a wide range of perfume compositions.

2.3.2 Application of the PQ2D® to Perfumery Quinary Systems
2.3.2.1 Effect of the Base Note on Perfume Formulation
The PQ2D® methodology can also be implemented for quinary perfumery systems in the same way that quaternary mixtures could be applied to the PTD® methodology, although with some graphical limitations. As it happened before, it is necessary to define pseudoquaternary compositions. For that purpose, the quaternary molar fractions

of the fragrant components are recalculated in a free basis of the fifth component, as follows:

$$x'_A = \frac{x_A}{x_A + x_B + x_C + x_S}, \quad x'_B = \frac{x_B}{x_A + x_B + x_C + x_S},$$

$$x'_C = \frac{x_C}{x_A + x_B + x_C + x_S}, \quad x'_S = \frac{x_S}{x_A + x_B + x_C + x_S} \tag{2.25}$$

where x'_i is a pseudoquaternary composition for each of the four fragrant components (A, B, C, S) in the quinary system (A, B, C, D, S).

To illustrate this application, we will evaluate the effect of introducing a fixative and the effect of incorporating water in a quaternary mixture. In the former case, the predicted quinary system consists of limonene (A), geraniol (B), vanillin (C), tonalide (D), and ethanol (S). Tonalide is considered at fixed compositions since it is only (strongly) perceived when nearly pure (see Figs 2.14 and 2.15). The ternary subsystems for the different quinary mixtures simulated are presented in Fig. 2.18, considering constant mole fractions of

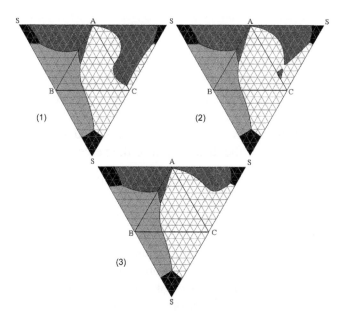

Fig. 2.18. Ternary subsystems for the three quinary perfume mixtures: (1) $x_{tonalide} = 0.10$, (2) $x_{tonalide} = 0.15$, (3) $x_{tonalide} = 0.20$. (A) Top note (dark gray, limonene); (B) middle note (light gray, geraniol); (C) base note (white, vanillin); (S) solvent (black, ethanol). Adapted with permission from Teixeira et al. (2009). © 2009, John Wiley & Sons.

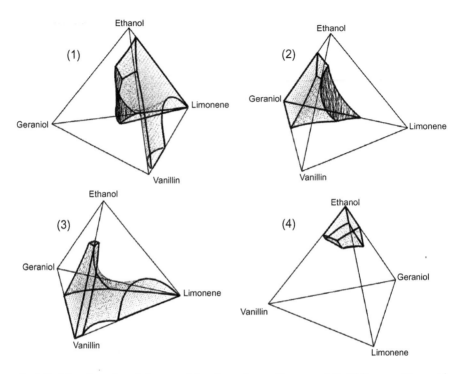

Fig. 2.19. Odor volumes for each fragrance of the quinary mixture with $x_{tonalide} = 0.10$: (1) limonene, (2) geraniol, (3) vanillin, (4) ethanol. Adapted with permission from Teixeira et al. (2009). © 2009, John Wiley & Sons.

tonalide: (1) $x_{\text{tonalide}} = 0.10$, (2) $x_{\text{tonalide}} = 0.15$, and (3) $x_{\text{tonalide}} = 0.20$, respectively. The behavior of the odor space for the whole quinary systems are presented in Figs 2.19–2.21 using the PQ2D® methodology. Once more each fragrance odor volume is presented separately in order to have a better view of the distribution of the perceived odor.

The incorporation of a second base note like tonalide in this perfumery system produces a clear effect in the odor space, revealing its fixative properties. Again, Fig. 2.18 shows the outside of the PQ2D® while Figs 2.19–2.21 reveals the whole PQ2Ds. As the tonalide concentration increases, there is a reduction of the odor volume for limonene because it tends to be more retained in the liquid. That is an evidence of the fixative effect of tonalide on the less polar molecules. On the opposite direction, the odor volume for vanillin increases with the concentration of tonalide, showing that vanillin tends to be "pushed out" of the solution and, thus, more intensely perceived.

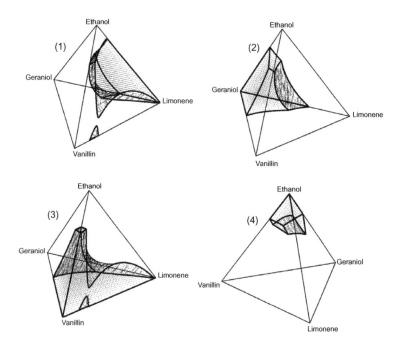

Fig. 2.20. Odor volumes for each fragrance of the quinary mixture with $x_{tonalide} = 0.15$: (1) limonene, (2) geraniol, (3) vanillin, (4) ethanol. Adapted with permission from Teixeira et al. (2009). © 2009, John Wiley & Sons.

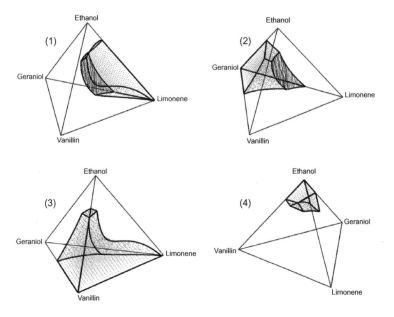

Fig. 2.21. Odor volumes for each fragrance of the quinary mixture with $x_{tonalide} = 0.20$: (1) limonene, (2) geraniol, (3) vanillin, (4) ethanol. Adapted with permission from Teixeira et al. (2009). © 2009, John Wiley & Sons.

2.3.2.2 Effect of Water on Perfume Formulation

Another important topic is the effect of water on perfume formulations. That can also be illustrated by means of the PQ2D® methodology. Water is a solvent commonly used in perfumery (especially in more "diluted" fragrances: *eau frâiche, eau de cologne*). Consider the addition of water to some of the quaternary systems already studied, using a constant composition ($x_{water} = 0.45$). The behavior of the quinary systems produced can be shown with the PQ2D®, in order to illustrate the influence of water in the odor space. The PTD® of the tetrahedron faces and the PQ2D® for the systems of limonene + geraniol + (vanillin or galaxolide) + ethanol + water are presented in Figs 2.22–2.24.

The introduction of water in the perfumery systems changes the shapes of the odor volumes for the different fragrance ingredients (compare Figs 2.22–2.24 with Figs 2.14–2.17–the corresponding systems without water). While for the mixture with galaxolide the changes are basically on the size of the odor volumes, the mixture with vanillin has clear changes also in their shape. Related to this effect is the fact that nonpolar limonene is pushed out of the solution more strongly in the presence of water, and so its odor volume is larger. Besides, the polar component ethanol is more retained in the solution with water, thus slowing down its evaporation rate and reducing its initial perception. Remarkably, these predictions are supported by experimental evidences from perfumery.

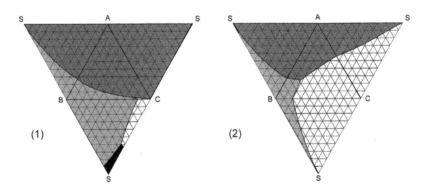

Fig. 2.22. Pseudoquaternary subsystems for the quinary perfume mixtures: (1) limonene + geraniol + vanillin + ethanol + water and (2) limonene + geraniol + galaxolide + ethanol + water. (A) Top note (dark gray, limonene); (B) middle note (light gray, geraniol); (C) base note (white, vanillin or galaxolide); (S) solvent (black, ethanol). Water composition was set equal to 45 mol% in both cases. Adapted with permission from Teixeira et al. (2010). © 2010, John Wiley & Sons.

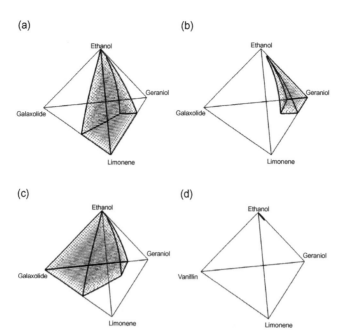

Fig. 2.23. Perfumery fragrance volumes for the quinary system (1) limonene + geraniol + vanillin + ethanol + water: (a) limonene, (b) geraniol, (c) vanillin, (d) ethanol. Water composition was set equal to 45 mol%. Adapted with permission from Teixeira et al. (2010). © 2010, John Wiley & Sons.

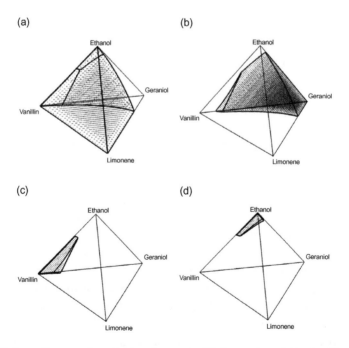

Fig. 2.24. Perfumery fragrance volumes for the quinary system (2) limonene + geraniol + galaxolide + ethanol + water: (a) limonene, (b) geraniol, (c) galaxolide, (d) ethanol. Water composition was set equal to 45 mol%. Adapted with permission from Teixeira et al. (2010). © 2010, John Wiley & Sons.

2.4 PERFUMERY OCTONARY SYSTEM

As previously seen, the developed PQ2D$^{®}$ methodology uses a graphic representation of a three-dimensional view using a tetrahedron to map the perceived odor of quaternary and quinary fragrance mixtures. However, if we want to introduce more fragrance ingredients, it becomes impossible to visualize it in its entirety: it would be necessary to map a multidimensional space (but plotting and understanding more than three dimensions on a plane paper is quite complex). Thus, although the methodology works for N component mixtures of fragrance ingredients (as well as the developed software), it is restricted by our own limitation of the visual perception. It is not possible to represent in a visual way (e.g., using a polyhedron) higher dimensional systems in their entirety, once Eq. (2.20) has to be satisfied. One way to tackle this problem would be by performing projections and cuts to reduce dimensionality. This visualization approach was already used in high-dimensional liquid–liquid equilibrium phase diagrams using the Jänecke or normalization projection, the parallel or lumping projection, the Cruickshank projection, and different isobaric and isothermal cuts (Harjo et al., 2004). Despite being valuable for reducing dimensionality, some of these projections continue to be difficult to visualize. Thus, the work to extend the graphic tools (PTD$^{®}$, PQ2D$^{®}$) to multicomponent mixtures is still running.

With certain limitations, the PTD$^{®}$ and PQ2D$^{®}$ can be used to explore the odor character of multicomponent mixtures. As an example, consider an octonary perfume system. This perfume mixture will include six fragrance ingredients (limonene + α-pinene + geraniol + linalool + vanillin + galaxolide) and two solvents (ethanol + water). Such perfumery system consists of two top notes, two middle notes, and two base notes in a binary solvent matrix. In order to represent the headspace odor of this octonary mixture in the PQ2D$^{®}$, a constraint has to be introduced in the chemical compositions: for example, two components of the mixture can be located together in each vertex of the tetrahedron (two top notes, two middle notes, two base notes, and two solvents). Thus, each vertex represents a type of component (top, middle, and base notes and solvent). But to do that, a specific ratio of these two components must be kept. In this case, two different component ratios were considered: one for the fragrance ingredients and another for the solvents, as given in Eqs (2.26) and (2.27):

$$x_j = \frac{0.55}{0.45} x_i \qquad (2.26)$$

$$x_{S_2} = \frac{0.60}{0.40} x_{S_1} \qquad (2.27)$$

where i and j correspond to each fragrance couple placed in a vertex of the tetrahedron, while s_1 and s_2 correspond to ethanol and water, respectively. Following this reasoning, in each vertex of the tetrahedron there will be a binary mixture with a specific ratio of compositions, and across each edge there will be a quaternary mixture where every two components have a composition that is determined by its couple.

The odor character of the subsystems composing the octonary perfume mixture is shown in Fig. 2.25 for the different PTDs of the tetrahedron diagram. The whole behavior of the octonary perfume system is presented in Fig. 2.26 by using the PQ2D® with the odor volumes of each component. Some of the fragrance ingredients and solvents have no representative odor volume for this mixture once they

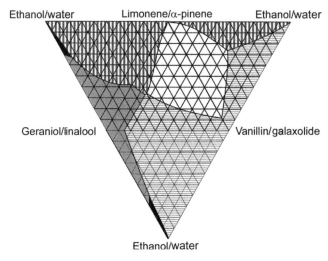

Fig. 2.25. *Subsystems composing the octonary perfume mixture. The figures are the projection of the tetrahedron faces. Limonene: no odor; α-pinene: green vertical stripes; geraniol: light gray; linalool: no odor; vanillin: white; galaxolide: yellow horizontal stripes; ethanol: black; water: no odor.* Adapted with permission from Teixeira (2011). © 2011, M.A. Teixeira.

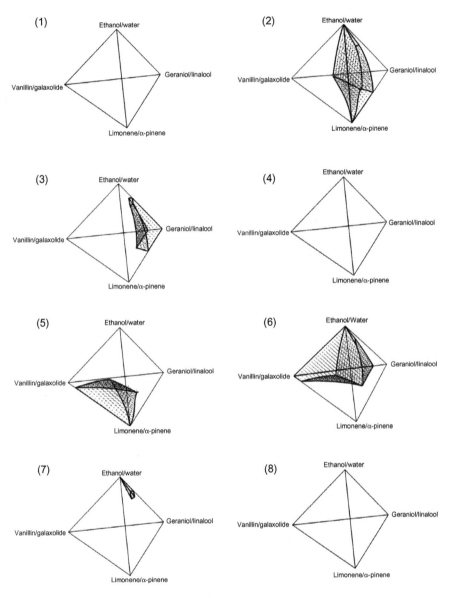

Fig. 2.26. Perfumery fragrance volumes for the octonary perfume system: (1) limonene, (2) α-pinene, (3) geraniol, (4) linalool, (5) vanillin, (6) galaxolide, (7) ethanol, and (8) water. Adapted with permission from Teixeira (2011). © 2011, M.A. Teixeira.

are not more intensely perceived than the other fragrances. Nevertheless, they may contribute to the odor as a nuance or a secondary odor. Moreover, there will not be an odor volume for water since it is odorless. From the predicted results obtained, it is possible to draw some conclusions: (i) in a multicomponent mixture not all fragrance components will be strongly perceptible; (ii) the combination of top notes may lead to a stronger initial perception of only one of those (depending on the composition) which explains the fact that limonene has no odor volume; (iii) base notes play an important role in perfumery not only for the fixative and long lastingness effects but also because their smell may be perceived from the beginning as background odors (odor volumes of vanillin and galaxolide); and (iv) a stronger perception of the fragrance ingredients reduces the range of compositions (odor volume) where ethanol is strongly perceived. This is important because its smell is undesired for the final perfume product.

2.5 CONCLUSION

The PTD® and PQ2D® tools allow presenting in a simple and easy-to-interpret graph the odor space of perfume mixtures. They can be used to explore the starting formulation of the perfume mixture, or to compare the effect of different components: here we have shown the influence of base notes and the role of water in the impact of the fragrance. The methodology also allows showing the qualitative odor character of these mixtures. It uses the concept of the OV, which is defined as the ratio between the concentration of an odorant in the gas phase and its corresponding odor threshold value, as previously defined. But the methodology is not restricted to this odor intensity model due to its modular conception.

A detailed study to evaluate the effect of nonidealities in the liquid mixture was also performed. For that purpose, the UNIFAC method was considered for the prediction of activity coefficients in the liquid phase (γ_i). As a result, it was shown that the perception of fragrances is not straightforward, and looking solely at the physicochemical properties of a single component isolated from the mixture may lead to erroneous conclusions. Rather, it must include the properties of all fragrance ingredients mixed in the perfume. Finally, the methodology was experimentally validated using a ternary mixture of limonene, geraniol, and vanillin with and without ethanol as a solvent. It was obtained a good agreement between the predicted OVs and those experimentally measured for the case of limonene and geraniol, though for vanillin they were poorly estimated (Gomes et al., 2008). In this way, further improvements are necessary to refine this tool and accurately predict the perceived odor of mixtures: either from thermodynamic models or from a better understanding of odor perception. Despite these methodologies will not replace the unparalleled olfactory ability of perfumers, they may help their work on the path for the design of new fragrances.

REFERENCES

Abrams, D., Prausnitz, J. M., Statistical Thermodynamics of liquid mixtures: A new expression for the excess Gibbs energy of partly or completely miscible systems, *AIChE Journal*, 21: 116–128, 1975.

AIHA, Odor Thresholds for Chemicals with Established Occupational Health Standards, Fairfax, VA, American Industrial Hygiene Association, TRC—Environmental Consultants, 1989.

Balk, F., Ford, R., Environmental risk assessment for the polycyclic musks AHTN and HHCB in the EU—I. fate and exposure assessment, *Toxicology Letters*, 111: 57–79, 1999.

Behan, J. M., Perring, K. D., Perfume Interactions with sodium dodecyl-sulfate solutions, *International Journal of Cosmetic Science*, 9 (6): 261–268, 1987.

Cain, W. S., Schiet, F. T., Olsson, M. J., de Wijk, R. A., Comparison of models of odor interaction, *Chemical Senses*, 20: 625–637, 1995.

Cain, W. S., Schmidt, R., Can we trust odor databases? example of t- and n-butyl acetate, *Atmospheric Environment*, 43 (16): 2591–2601, 2009.

Calkin, R., Jellinek, S., Perfumery: Practice and Principles, New York, NY, John Wiley & Sons, 1994.

Carles, J., A method of creation in perfumery, *Soap Perfumery and Cosmetics*, 35: 328–335, 1962.

Chapman, J., MATLAB Programming for Engineers, California, Brooks Cole Publishing Company, 2000.

Chapra, S., Canale, R., Numerical Methods for Engineers: With Software and Programming Applications, Boston, MA, McGraw-Hill, 2002.

Curtis, T., Williams, D. G., Introduction to Perfumery, New York, NY, Ellis Horwood, 1994.

Devos, M., Rouault, J., Laffort, P., Standardized Olfactory Power Law Exponents, Dijon, France, Editions Universitaires Sciences, 2002.

Fráter, G., Müller, U., Kraft, P., Preparation and olfactory characterization of the enantiomerically pure isomers of the perfumery synthetic galaxolide, *Helvetica Chimica Acta*, 82: 1656–1665, 1999.

Fredenslund, A., Jones, R., Prausnitz, J. M., Group-contribution estimation of activity coefficients in nonideal liquid mixtures, *AIChE Journal*, 21 (6): 1086–1099, 1975.

Gomes, P. B., Mata, V. G., Rodrigues, A. E., Experimental validation of perfumery ternary diagram methodology, *AIChE Journal*, 54 (1): 310–320, 2008.

Gubbins, E., Applications of Molecular Theory to Phase Equilibrium Predictions, Models for Thermodynamic and Phase Equilibrium Calculations, Sandler, NY, Marcel Dekker, Inc., 1994.

Harjo, B., Ng, K. M., Wibowo, C., Visualization of high-dimensional liquid–liquid equilibrium phase diagrams, *Industrial & Engineering Chemistry Research*, 43: 3566–3576, 2004.

Hau, K. M., Connell, D. W., Quantitative structure–activity relationships (QSARs) for odor thresholds of volatile organic compounds (VOCs), *Indoor Air-International Journal of Indoor Air Quality and Climate*, 8 (1): 23–33, 1998.

Klamt, A., COSMO-RS: From Quantum Chemistry to Fluid Phase Thermodynamics and Drug Design, Amsterdam, Elsevier, 2005.

Laffort, P., Dravnieks, A., Several models of suprathreshold quantitative olfactory interaction in humans applied to binary, ternary and quaternary mixtures, *Chemical Senses*, 7 (2): 153–174, 1982.

Leffingwell, J. C., Leffingwell, D., Odor Thresholds, from <http://www.leffingwell.com/odorthre.htm,> Last Accessed on May 2012.

Mata, V. G., Gomes, P. B., Rodrigues, A. E., Science Behind Perfume Design, Second European Symposium on Product Technology (Product Design and Technology), Groningen, The Netherlands, 2004.

Mata, V. G., Gomes, P. B., Rodrigues, A. E., Effect of nonidealities in perfume mixtures using the perfumery ternary diagrams (PTD) concept, *Industrial & Engineering Chemistry Research*, 44 (12): 4435–4441, 2005a.

Mata, V. G., Gomes, P. B., Rodrigues, A. E., Engineering perfumes, *AICHE Journal*, 51 (10): 2834–2852, 2005b.

Mata, V. G., Gomes, P. B., Rodrigues, A. E., Perfumery ternary diagrams (PTD): A new concept applied to the optimization of perfume compositions, *Flavour and Fragrance Journal*, 20 (5): 465–471, 2005c.

Mata, V. G., Rodrigues, A. E., A new methodology for the definition of odor zones in perfumery ternary diagrams, *AIChE Journal*, 52 (8): 2938–2948, 2006.

MathWorks, Partial Differential Equation Toolbox—ComsoLab MATLAB's Users Guide, 2002.

Mayer, V. A., Fazio-Fluehr, P. C., Arendt, S. A., ASTM Dictionary of Engineering Science and Technology, Mayfield, PA, ASTM International, 2005.

Müller, P., Neuner-Jehle, N., Etzweller, F., What makes a fragrance substantive? *Perfumer & Flavorist*, 18: 45–49, 1993.

Nagata, Y., Measurement of odor threshold by triangle odor bag method. odor measurement review, *Office of Odor*, 118–127, 2003.

Nocedal, J., Wright, S., Numerical Optimization, New York, NY, Springer, 1999.

Ohloff, G., Pickenhagen, W., Kraft, P., Scent and Chemistry—The Molecular World of Odors, Zurich, Wiley-VCH, 2012.

Panagiotopoulos, A. Z., Molecular simulation of phase-equilibria—simple, ionic and polymeric fluids, *Fluid Phase Equilibria*, 76: 97–112, 1992.

Patte, F., Etcheto, M., Laffort, P., Selected and standardized values of suprathreshold odor intensities for 100 substances, *Chemical Senses & Flavour*, 1 (3): 283–305, 1975.

Perry, R. H., Perry's Chemical Engineers' Handbook, USA, McGraw-Hill, 1997.

Poling, B., Prausnitz, J. M., O'Connell, J., The Properties of Gases and Liquids, McGraw-Hill, 2004.

Renon, H., Prausnitz, J., Local compositions in thermodynamic excess functions for liquid mixtures, *AIChE Journal*, 14: 135–144, 1968.

Rodríguez, O., Teixeira, M. A., Rodrigues, A. E., Prediction of odor detection thresholds using partition coefficients, *Flavour & Fragrance Journal*, 26, (6): 421–428, 2011.

Stevens, S. S., On the psychological law, *The Psychological Review*, 64 (3): 153–181, 1957.

Teixeira, M. A., Perfume Performance and Classification: Perfumery Quaternary–Quinary Diagram (PQ2D®) and Perfumery Radar. Dept. of Chemical Engineering, Faculty of Engineering of University of Porto, Phd Thesis, 2011.

Teixeira, M. A., Rodríguez, O., Mata, V. G., Rodrigues, A. E., Perfumery quaternary diagrams for engineering perfumes, *AIChE Journal*, 55 (8): 2171–2185, 2009.

Teixeira, M. A., Rodríguez, O., Rodrigues, A. E., The perception of fragrance mixtures: a comparison of odor intensity models, *AIChE Journal*, 56 (4): 1090–1106, 2010.

Teixeira, M. A., Rodríguez, O., Rodrigues, A. E., Odor Detection and Perception: An Engineering Perspective Chapter 1 Advances in Environmental Research, 14, USA, Nova Publishers, 2011a.

TGSC, The Good Scents Company, from <http://www.thegoodscentscompany.com/>, Last Accessed on February 2012.

Tochigi, K., Tiegs, D., Gmehling, J., Kojima, K., Determination of new ASOG parameters, *Journal of Chemical Engineering Japan*, 23 (4): 453–463, 1990.

van Gemert, L. J., Compilations of Odour Threshold Values in Air, Water and Other Media, the Netherlands, Oliemans Punter & Partners BV, 2003.

Walas, S., Phase Equilibria in Chemical Engineering, Boston, MA, Butterworth–Heinemann, 1985.

Wei, N., The 3-dimensional phase diagram in quaternary systems of polymers and solvents, *Journal of Applied Polymer Science*, 28: 2755–2766, 1983.

Zwislocki, J. J., Sensory Neuroscience: Four Laws of Psychophysics, New York, NY, Springer Science + Business Media, 2009.

Performance of Perfumes

In this chapter, we will address fragrance performance, a very relevant topic for the design of novel and improved fragranced products. Fragrance performance is often assessed by experimental sensory evaluation using panelists, not so by predictive models. Consequently, it introduces human interpersonal variability and increases the time needed for such evaluations. It is no surprise that it would be a tremendous breakthrough to have theoretical models capable of doing so, in seconds.

In order to account for fragrance performance, one should take into account three important factors: first, the release of fragrant molecules from a perfume, then their propagation in the air, and, finally, their perception at the olfactory level. These three steps, though, bring a series of questions: How can we model fragrance release from a solution? As a proof of concept, will it be feasible to experimentally measure and validate fragrance evaporation? What will be the relevance of such analysis for fragrance design and performance? Ultimately, how can we quantify a qualitative property like fragrance performance?

In order to answer some of these questions, we have selected four performance parameters commonly used in perfumery: impact, diffusion, tenacity, and volume. We have modeled the performance of selected fragrance mixtures using these parameters and the results obtained were compared to experimental evaluations.

The diffusion of odorant mixtures in air was addressed together with the assessment of their performance in terms of olfactory perception. A predictive model to account for the diffusion of fragrances in air was applied to several mixtures of fragrance ingredients and their performance using perfumery concepts was quantified. The Perfumery Ternary Diagram (PTD®) and the Perfumery Quaternary-Quinary Diagram (PQ2D®) methodologies were recalled here for the representation of perfume evaporation paths using evaporation lines. The performance of these perfume mixtures was also analyzed considering the odor intensity and lastingness of the odorants at different distances from the point of application of the perfume and over time. For quantifying the evaporation of fragrances, several predictive G^E methods based on the group-contribution concept were tested for the prediction of multicomponent vapor–liquid equilibria (VLE). Experimental data of equilibrium compositions was measured by headspace gas chromatography (HS-GC) to validate this step as well.

3.1 FRAGRANCE PERFORMANCE

Fragrance performance is often evaluated empirically through olfactory evaluations performed either by nontrained panelists or by perfumers. Some key performance parameters have been defined by the industry, in an attempt to compare the olfactory effects of different fragrances or different formulations. Among these key performance parameters there are four that account for the effect of time and distance from the source of a perfume. These four performance parameters are the so-called impact, diffusion, tenacity, and volume (Cortez-Pereira et al., 2009; Teixeira et al., 2009a):

1. *Impact* is an immediate olfactory sensation and is a measure of the intensity of a perfume in the first moments after an application. An example is when sniffing a perfume from a blotter or right after its application onto the skin.

2. *Tenacity* is the ability of a fragranced mixture to retain its characteristic odor during the so-called dry-down stage. It is a performance index that measures the persistence of a fragrance for long times after its application but near the evaporating source, like how long the perfume lasts in the skin after applying it.
3. *Diffusion* refers to the efficacy of a perfume at some distance from the source, representing how fast a fragrance radiates in space and permeates into the surrounding environment.
4. *Volume* is the effectiveness of a perfume over distance, some time after application.

These odor performance parameters are time and distance dependent. Together they provide a picture of the propagation of a given perfume or fragrance in air, allowing to compare the effectiveness of different fragrance ingredients and formulations. The four parameters can be easily understood through Fig. 3.1.

Although the definition of fragrance performance by words may look simple, it is complex to measure and model it (and ultimately predict it!). This is so because it is a function of complex properties: the selected fragrance ingredients, their composition, selection of matrix or support of application, as well as the release and propagation rates of the different odorants, and, finally, the effect of their concentration on

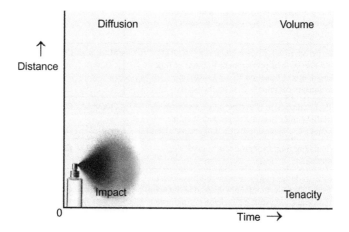

Fig. 3.1. Parameters used by perfumers to evaluate the performance of a perfume as a function of the intensity and character of the perceived odor with time and distance. Adapted with permission from Teixeira (2011). © 2011, M.A. Teixeira.

the odor intensity and character as perceived by the human nose. Thus, despite the fact that product performance remains indisputably a very important issue for companies (and those from F&F are not different), it is also complex to evaluate it. Ultimately, the target will always be the development of best seller fragrances or fragranced products, though the path to reach those is still cloudy and risky.

Both R&D departments from industry and academy have been studying different topics within the performance of fragrances. An overview of some relevant works is presented in Table 3.1. Note, however, that the term performance is commonly used in perfumery as a measure of the long lastingness or strength of a perfume/fragrance as

Table 3.1. Summary of Some Relevant Literature on Performance of Fragrances	
Approach	**References**
Measurement of fragrance intensity and base odor above skin using panelists and a labeled magnitude scale (LMS) for performance evaluation	Cortez-Pereira et al. (2009)
Evaluation of perfume performance based on chemical structure related properties and physicochemical properties of the chemicals together with calculation of the perceived odor intensity	Stora et al. (2001)
Combination of headspace measurements with olfactometry data and correlation of physicochemical properties for evaluation of fragrance performance	Gygax et al. (2001)
Definition of performance parameters like impact, tenacity, diffusion, volume, and substantivity as well as important properties for the evaluation of performance like the OV, odor threshold, or log P	Calkin and Jellinek (1994)
Prediction of the perceived odor of mixtures using the PTD methodology and also of their performance in terms of time and distance using a model based on Fick's Laws and performance parameters commonly used by the industry	Mata et al. (2005) Gomes, P. B. (2005)
Proposal of perfume compositions comprising nonsubstantive fragrance materials in order to enhance product shelf life, delivery effectiveness, and substantivity on different substrates	Duprey et al. (2010)
Methods of formulating fragrance products to mask the malodor of ammonia	Fadel et al. (2009)
Perfume compositions designed for use in wash-off systems to provide a high initial bloom with minimal residual perfume on the targeted system and a long sustained release of fragrance	Fadel et al. (2008)
Methods and compositions of perfume mixtures to improve the release of fragrance materials from an entrapment structure on a surface over time	Heltovics et al. (2004)
Source: Adapted with permission from Teixeira (2011). © 2011, M.A. Teixeira.	

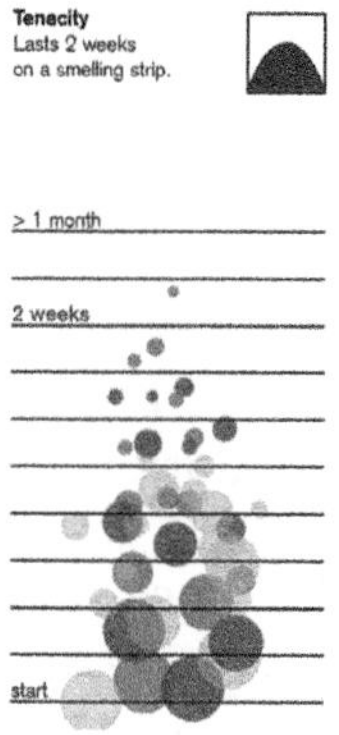

Tenacity
Lasts 2 weeks
on a smelling strip.

> 1 month

2 weeks

start

STABILITY & PERFORMANCE

RECOMMENDED APPLICATIONS	STABILITY	TYPICAL % USE	SUBSTANTIVITY & REMARKS	
fine fragrances	■■■■	1 - 10%	Heart and base note	
shampoo	■■■■	1 - 5%	wet: ■■■■	dry: ■■■■
shower gel	■■■	1 - 5%	bloom: ■■■	
soap	■■■	0.5 - 5%	foam: ■■■■	dry hand: ■■■■
detergent	■■■	0.5 - 5%	wet: ■■■	dry: ■■■
softener	■■■■	0.5 - 5%	wet: ■■■■	dry: ■■■■
APC	■■■	0.5 - 5%	–	
candle	■■■	0.5 - 5%	cold wax: ■■■	burning: ■■■

Fig. 3.2. Typical evaluation compendium for fragrance ingredients (e.g., Habanolide®) from Firmenich. Reproduced with permission by courtesy of Firmenich Ingredients Division, available at http://www.firmenich.com/e-catalog/index.lbl. © 2012 Firmenich.

shown in Fig. 3.2. These terms are generally related to the persistence of a fragrance on a paper blotter over time, although such an evaluation may seem too simplistic (it is frequently observed a completely different behavior of the fragrance when applied onto a textile or the human skin). In fact, within the perfumery business, the sensory evaluation of a pure fragrance ingredient or a formulated mixture is primarily based on the quantification of human responses to physicochemical stimuli. This process encompasses olfactory evaluations often carried out by perfumers but also by nontrained panelists as well. Of course, this methodology involves using human noses as measuring tools, either for a quantitative or for a qualitative and descriptive analysis (Cortez-Pereira et al., 2009).

According to Calkin and Jellinek (1994), the performance of a perfume is expressed by its ability to become noticeable. Thus, the performance of a perfumed product starts in the optimization of its composition, in order to obtain the maximum desired odor effect at the lowest possible concentration. Yet, several questions still remain to be unfolded: does the performance of a perfume depends on the intrinsic performance of its constituents? Does the art of perfumery involved in the formulation of the product also play a role in the performance of the product? How should the performance of a perfumed product be measured and, better yet, predicted?

In order to obtain a more precise and scientific perspective of how a perfumed product will be perceived, one should account not only for the time variable but also for space (or distance from the source of application of the perfume). In this way, the perfume is evaporating and diffusing through the surrounding air over time and distance and will be perceived differently depending on where we stand on these two variables. For that purpose, the four performance parameters defined above can be used and evaluated in order to define the lowest possible perfume dosage that allows the desired maximum odor effect (Cortez-Pereira et al., 2009; Teixeira et al., 2009a). There are some other performance parameters that may be useful. The odor life of a fragrance measures how long a fragrance note persists in the gas phase above a liquid mixture (headspace) with intensity higher enough to be smelled (e.g., with a concentration above its odor threshold value). Another parameter that must be referred here is the fragrance substantivity: it deals with the tendency of an essential oil or fragrance ingredient, when applied as a diluted aqueous solution, to bind itself to a solid surface like the skin. This property measures the adhesion of a fragrance to the substrate where it is applied, like its affinity to stay on a surface when this is moistened (Calkin and Jellinek, 1994; Cortez-Pereira et al., 2009). Some of the aforementioned odor performance parameters can be evaluated quantitatively or qualitatively using different experimental techniques or models that will be discussed later in this chapter.

3.2 THE RELEVANCE OF THE SELECTION OF FRAGRANCE INGREDIENTS

Having in mind the large number of available fragrances in the market (in the order of several thousands), it becomes easy to understand that the possible spectra of combinations for these ingredients, which are placed at the hands of a perfumer, are almost endless. This fact presents two direct implications: first, it allows the perfumer to expand all his/her artistic creativity in the formulation of new fragranced products with fascinating organoleptic properties. But on the other hand, it increases the entropy of such perfumery systems, especially at the molecular level (where the presence of a chemical base or a slight change in the selected composition may have undesirable effects). Consequently, this last issue brings an additional difficulty in designing new fragrances with improved performance and likeability for customers.

A parenthesis is made here to highlight that the F&F industry is thoroughly regulated by international organizations. These main organizations are the Research Institute for Fragrance Materials (RIFM), the International Fragrance Association (IFRA), and the REACH (Registration, Evaluation, Authorization, and Restriction of Chemical substances). These perform evaluations on chemical properties, their use and application, and collect data from the published scientific literature or suppliers' reports, thus defining a "Code of Practice" that may be accompanied by detailed recommendations (when appropriate), which can be used as voluntary guidelines by the industry to restrict (or even forbid) the use of chemicals that are believed to be potentially harmful for the human being. Additionally, some fragrance companies also apply in-house "black lists," thus imposing an even more stringent restriction on the raw materials that their perfumers are allowed to use (Abbe, 2000). Consequently, the imposition of such restrictions causes a reduction in the number of available fragrances for the formulation of new products. This complicates, even more severely, the job for perfumers who are willing to obtain fragrances with a desired olfactive target but simultaneously guarantee the stability and safety of the final product.

Thus, the formulation of fragranced products (either fine fragrances or functional perfumes like soaps, household cleaners, or detergents) is a complex, slow, and costly trial-and-error process that involves testing hundreds of different samples for the perceived odor (Calkin and Jellinek, 1994). Strikingly, as if all the above were not enough to make this job difficult, fragrances' behavior in terms of release (and propagation as well) may be completely diverse when incorporated in different matrices or supports. It is known that the perceived odor released from such products may be completely different. That has to do with the ratio of fragrance to solvent concentration, the types of fragrance raw materials selected, and the interactions occurring at molecular level between the fragrance ingredients and product bases (Behan and Perring, 1987; Teixeira et al., 2009b, 2011). Such fact complicates, even more, achieving a desired olfactive target and so different perceptions of fragrances' intensity and/or character can be found.

Apart from that, the characteristic odor elicited from an essential oil or a perfume may be due to a single ingredient with a major composition or to a trace component that evaporates rapidly and has a low odor detection threshold (Ryan et al., 2008). Anyway, in most cases the characteristic odor evaporating from a mixture is the result of a multicomponent combination of different odors.

From the physicochemical point of view, several variables play a role on the release of fragrance chemicals from mixtures (or from the skin, a textile, or a paper blotter): the saturated vapor pressure of each component, the octanol–water partition coefficient (or any other suitable measurement of polarity), the diffusion coefficient, the boiling point, the solubility in water, or even the molecular weight. In fact, the search for highly blooming odorants is also relevant for fragranced products like detergents and shampoos, which generally are applied in water dilution conditions. For such cases, for example, it is considered that the physical properties of the fragrant molecules may play a role on the perfume burst so that odorants with a partition coefficient of at least 3.0 and a boiling point of less than 260°C may be considered as having superior release properties (Fadel et al., 2008). Nonetheless, it is obvious that this is a simplified perspective of the problem because even within an emulsion system (as these cases apply), molecular interactions are occurring and may influence the release mechanism.

Independently of the pathway followed, the development of a fragrance product includes several issues that need to be addressed. So, from the point of view of the perception of fragrances incorporated into consumer products, the topics that must be taken into account are listed below: some are known and controlled (unmarked), other need to be better understood and controlled (marked with *), and, finally, there are other more problematic topics that are still in need of new technical routes and innovative approaches for their accurate and reproducible measurement (marked with [#]) (Kerschner, 2006).

1. odor threshold of each perfume ingredient,
2. amount of product used by the consumer,
3. mixing of fragrance ingredients,
4. fragrance evaporation from mixtures*,
5. minimum concentration used in the product*,
6. odor quality of the base*,
7. degree of interaction with the base*,

8. effect of processing the product[*],
9. fragrance perception by consumers[#],
10. stability of the perfume during storage[#],
11. interaction with the packaging[#],
12. amount of perfume deposited and retained after the wash[#],
13. rate of release over time and perfume performance[#].

For these reasons, F&F companies are investing capital and human resources in the design steps of their fragranced products for the development of experimental methodologies and predictive tools which, for example, may be capable of predicting VLE and dispersion of fragrance ingredients to evaluate and model fragrance performance.

3.3 EVAPORATION OF FRAGRANCE CHEMICALS

The process of dissipation of a fragrance from a specific matrix presents itself as an important step for the design and evaluation of novel fragrance materials or products containing fragrance ingredients. But it is also used as an instrument to aid in the risk assessment of fragranced products, namely for skin sensitization (Kasting and Saiyasombati, 2003). This is very importance because it is known that a wide range of factors influence the way we perceive a fragrance that is evaporating into the air. It is known that fragrance materials will behave differently if they are evaporating from a liquid mixture, a paper blotter, or the skin (where it varies from person to person as well). This is common problem that perfumers have to face in the preformulation stage of a fine fragrance (Burr, 2008).

Consequently, the evaporation of a fragrance mixture from a substrate or a base depends on their physicochemical properties, molecular interactions, temperature, or pH, just to mention a few. In this way, the study of all the phenomena involved in the formulation, behavior, release, diffusion, and perception of fragrances is of great importance. Depending on the type of product and its final application, different topics have to be circumvented by the R&D groups from the industries: (i) evaluation of both vapor–liquid or liquid–liquid (or even vapor–liquid–liquid) equilibrium of mixtures composed by fragrance ingredients at different temperatures may provide important information for fragrance development (Arce et al., 2002; de Doz et al., 2008); (ii) measurement of phase interactions for multicomponent mixtures composed by surfactants, water, and fragrance ingredients is also valuable for the industry (Tokuoka et al., 1994; Friberg, 2009; Friberg et al., 2009). As an example, Friberg et al. have thoroughly studied different physicochemical properties in fragrance evaporation from emulsion systems using ternary diagrams and algebraic calculations to withdraw the evaporation paths of fragrance chemicals (Friberg, 2009; Friberg et al., 2009); (iii) measurement of the solubility of fragrance materials in water and alcohols is of great importance to define the range of possible compositions where a perfume is a homogeneous solution and to evaluate skin disorders, since they are used in a number of cosmetic products (Domanska et al., 2008, 2010); (iv) studies involving humans for experimentation are also of relevance like the measurement of evaporation and absorption rates of fragrances from the skin (either *in vivo* or *in vitro*), and the development of theoretical models for these phenomena (Kasting and Saiyasombati, 2003). Furthermore, several studies concerning different topics within the evaporation and diffusion of fragrances released from different substrates or bases can be found in the literature. A nonexhaustive selection of reference works on these fields is presented in Table 3.2.

It should be highlighted the research of Friberg and coworkers on the mechanisms behind fragrance release from emulsion systems. In their approach, they combine phase diagrams with algebraic calculations to withdraw information from these ternary diagrams, as illustrated in Fig. 3.3. In this example, the evaporation paths of linalool emulsions were experimentally determined and compared to those calculated from vapor pressures, for a series of emulsions with different oil/water (O/W) ratios. In the ternary diagrams of Fig. 3.3 the three-phase regions are indicated by a triangle delimited by the three tie lines defining the composition of the aqueous liquid (Aq), the liquid crystal (LC), and the oil phase (Oil). This is relevant since most studies on this subject do not address the variation of the conditions during a prolonged evaporation (e.g., composition or the number and structures of the emulsion phases) which is experienced in most emulsion applications.

Through their methodology, it is possible to predict the evaporation path or the change in the vapor pressure during the evaporation process of fragrance chemicals from an oil/water emulsion. Other effects were also evaluated, to account for deviations from equilibrium conditions in the evaporation of emulsions like the weight of relative

Table 3.2. Summary of Relevant Previous Studies on Evaporation and Release of Fragrances

Approach	References
From the skin	
Measurement of the evaporation rates of fragrance ingredients alone and in multicomponent mixtures with and without fixatives using dynamic headspace	Vuilleumier et al. (1995)
Release of perfume from the skin and evaluation of the physical and chemical interactions between them, using headspace analysis with trap adsorbents	Behan et al. (1996)
Evaporation of fragrances using a skin substitute (*in vitro*) and *in vivo* measurements with headspace techniques	Schwarzenbach and Bertschi (2001)
Evaporation of fragrances from the human forearm under *in vivo* conditions considering evaporation and absorption rate constants and estimation of the release using physicochemical properties following nearly first-order kinetics. Measurement and modeling of the evaporation rates of fragrance ingredients in multicomponent mixtures with and without fixatives using dynamic headspace	Kasting and Saiyasombati (2001, 2003) Saiyasombati and Kasting (2004)
Measurement of fragrance intensity and base odor above skin using panelists and a LMS	Cortez-Pereira et al. (2009)
Evaluation of the effect of skin properties on the release of fragrances from the skin using dynamic headspace traps and the OV concept	Baydar et al. (1995)
Study of fragrance performance by measurement of vapor concentrations over time and evaluation of pleasantness of the aroma in the vicinity of the fragranced skin	Baydar et al. (1996)
Comparison of the vapor phase around a plant and the oil applied on the skin using solid phase-micro extraction (SPME) technology and study of the effect of skin properties on the release	Mookherjee et al. (1998)
From liquid mixtures	
Evaporation from ethanol–water solutions was measured in a wind tunnel using a circular cell and modeled with Gibbs adsorption equation	Speading et al. (1993)
Development of a methodology for the prediction of the perceived odor of ternary to quaternary mixtures using the PTD and the diffusion of fragrance mixtures in the air using the Fick's Law	Mata et al. (2005)
Prediction of the evaporation and diffusion of fragrance mixtures in the air using VLE methods and a model based on the Fick's Law	Teixeira et al. (2009a)
Measurement of vapor compositions in equilibrium with the liquid phase of fragrance mixtures and comparison with different VLE predictive methods and human olfactory evaluations	Teixeira et al. (2011)

(Continued)

Table 3.2. (Continued)	
Approach	**References**
From emulsions	
Calculation and measurement of evaporation paths of several fragrances in oil emulsions using phase diagrams and modeling by algebraic methods using physicochemical properties. Different emulsion systems were evaluated mostly of the type surfactant−fragrance−water. Different properties of these systems with influence on the evaporation rate were evaluated	Aikens et al. (2000); Friberg (1998, 2007); Friberg et al. (2010)
Evaluation of the hypothesis that odor intensity is controlled by the concentration of the aqueous phase of an oil-in-water emulsion using the headspace technique	Brossard et al. (2002)
Measurement of the release of fragrances (and study of the presence of ethanol) from amphiphilic multiarm star-block copolymers using thermogravimetry and dynamic headspace analysis	Ternat et al. (2008)
Other applications	
Evaluation of the substantivity of fragrances applied in laundered and dry-out fabrics. Correlation of the affinities of fragrances in standard fabric softener and detergent solutions using the octanol−water partition coefficient	Escher and Oliveros (1994)
Evaluation of the deposition of fragrances on textiles and their evaporation using perfumed fabric softeners was measured in a headspace cell	Stora et al. (2001)
Olfactory evaluation of the odor intensity of different types of fabrics finished with cyclodextrins and impregnated with fragrances was performed for over a year using OVs	Martel et al. (2002)
Microencapsulation of fragrances for industrial application on fabrics and evaluation of the release upon abrasion and washing cycles	Rodrigues et al. (2009)
Impregnation of microcapsules containing fragrances in fabrics and evaluation of the released fragrance before and after washing using an electronic nose	Specos et al. (2010)
Online measurement of fragrance release from cotton towels impregnated with microcapsules using a cryogenic system for headspace analysis	Haefliger et al. (2010)
Measurement and prediction of the evaporation of fragrances from microencapsulated textiles, their diffusion, and perception in the surrounding air	Teixeira et al. (2011)
Adapted with permission from Teixeira (2011). © 2011, M.A. Teixeira.	

humidity on the evaporation rate or the growth and reduction of phase volumes at nonequilibrium conditions (Bozeya et al., 2009).

On the other hand, the evaporation of fragrance materials involving clinical trials such as the release from the skin (e.g., forearm) is also of great relevance although experiments of *in vivo* release of fragrant

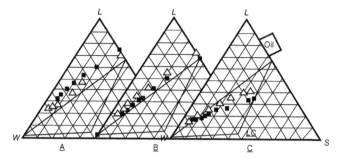

Fig. 3.3. Evaporation paths for three emulsions with different initial compositions (A, B, C) for the ternary system of water (W), linalool (L), and surfactant laureth-4 (S). Open triangles represent experimental values and black squares are calculated compositions. Adapted with permission from Al-Bawab et al. (2009). © 2009, John Wiley & Sons.

chemicals are scarce in the literature. In this way, the volatiles released from the skin could be identified and quantified, and then the odor intensity could be calculated over time, using for example the odor value (OV). It is also possible to obtain valuable information from the literature about parameters influencing the evaporation process from the skin: the effect of physical properties of the skin, the mechanism of absorption into the skin, or the kinetics of the release, among others. It is known that the skin interacts with fragrance chemicals and appears to substantially slow down their evaporation (especially with oily skin, see for example (Baydar et al., 1995) for a comparison with free evaporation from a glazed surface). This behavior has to do with complex fragrance–skin interactions. The skin effect on each component of a fragrance mixture varies depending on its nature (e.g., polarity, molecular size, among others). Within this issue, it would be reasonable to expect that moderately hydrophobic ingredients would present higher affinity for the oil phase on the skin, and thus be held more strongly than others. Conversely, the presence of perspiration may drive off very hydrophobic materials such as terpenes (Behan and Perring, 1987). Another consideration is that the skin, like any other biological tissue, may act as a reservoir for fragrance chemicals which can be absorbed and, consequently provoke skin sensitization. To evaluate the balance between percutaneous penetration and evaporative losses of fragrance from the skin, the ratio of the chemical–physical properties of vapor pressure to the product of its octanol–water partition coefficient and its aqueous solubility can be calculated to give some reasonable approximation of the phenomena (Guy, 2010).

Finally, it is also important to address the application of fragrance materials on fabrics and evaluate their release and performance in air. There are several studies on the development of polymeric microcapsules containing fragrances inside for impregnation into fabrics and posterior evaluation of the released volatiles after dry washing or abrasion cycles using HS-GC, cryogenic systems coupled with chromatography, electronic noses, or olfactory evaluations, as shown in Table 3.2. In this way, depending on the type of product (e.g., liquid soap, detergent, and bath oil) in which a fragrance is incorporated (which also influences the selected concentration) it is not surprising to observe a variation in the vapor composition that will change fragrance intensity and/or character as perceived by the nose. The primary explanation would arise from the dilution process itself, which has to do with the psychophysical properties of the fragrance and the solvent (e.g., saturated vapor pressure and molecular weight). Yet, for different product bases which themselves have very weak odors, it is seen that using the same perfume dilution results in a significantly different odor perception. For that reason, molecular interactions within a product play an important role in the evaporation of the fragrance ingredients.

3.4 DIFFUSION OF FRAGRANCES

Whether it is a detergent, a shampoo, or an air freshener, the performance of such products is measured not only by their efficiency for the function they were designed for but also by the fragrance intensity and quality released during application. Thus, the evaluation of fragrance diffusion becomes a central topic to assess and quantify fragrance performance. Although this is not deeply investigated in terms of perfumery, there is no doubt that fragrance dispersion will influence consumer likeability and, consequently, product sales.

The description of diffusion processes involves the use of mathematical models based on some fundamental law. For such a law, there are essentially two common choices: the simplest is the Fick's Law which makes use of a diffusion coefficient parameter and is commonly used to describe diffusion processes alone. On the other hand, there is a different approach that involves the use of a mass transfer coefficient (Bird et al., 1960, 2002; Cussler, 2007). The preference of one approach over the other, according to Cussler (2007), is the result of a *compromise between ambition and experimental resources.* If it is desirable to follow a more fundamental perspective, diffusion coefficients should be preferred. But if a more approximate and phenomenological experiment is planned, then such approximation would lead to mass transfer coefficients. From a Product Engineering perspective, it is important to have a predictive model for fragrance diffusion. Thus, the diffusion coefficients' approach has been followed.

3.4.1 Perfume Diffusion Model

Let us consider the simplest problem of fragrance diffusion, allowing only one direction for the process. A liquid fragrance ingredient or a perfume mixture will be evaporating into the air above. Vapor molecules will, then, start to diffuse and propagate in the surroundings and will eventually reach our nose to be perceived with some intensity and recognized with some olfactory quality. In order to model that we need a diffusion model capable of simulating the propagation of a liquid perfume mixture over time (t) and distance (z). This physical system is presented in Fig. 3.4 and consists of a very small volume of liquid perfume that is evaporating in the air over time and diffusing upward through the gas phase (air) above it. The liquid mixture is considered nonideal (for mixtures), with a small volume as it typically happens when we spray perfume on our body. Due to the low volume of perfume considered in this study (assumed to be 1 mmol $\approx 50-100\ \mu L$), it is expected that mass transfer resistances in the liquid phase can be neglected. This assumption is based on the following:

1. The methodology is being applied to perfume utilization, that is, when a perfume is sprayed on the skin or clothes.

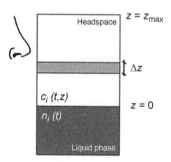

Fig. 3.4. Scheme of the simulated liquid and gas phases of the perfume system. Adapted with permission from Teixeira et al. (2009a). © 2009, Elsevier.

2. It was considered an equivalent area of the liquid–gas interface of 0.071 m² (∼0.3 m of diameter), which gives a liquid thickness of approximately 10^{-6} m. Under these conditions, we are well below the film thickness usually considered for mass transfer resistance in the film theory, 10^{-4} to 10^{-5} m for liquids (Taylor and Krishna, 1993).

Thus, mass transfer resistance in the liquid phase was neglected and the liquid was assumed as perfectly mixed. Consequently, we have followed the approach of Fick for describing the diffusion of fragrances in air.

Moreover, it is expected that both volume (V) and composition (x_i) of the liquid mixture will change with time as the perfume evaporates. The composition of the liquid mixture is defined by the number of moles of each intrinsic component (n_i) while the composition of the gas phase is determined by the molar concentration (c_i) of each fragrance ingredient. Throughout this study, both pressure (P) and temperature (T) are assumed to be kept constant while evaporation takes place.

3.4.1.1 Gas Phase

It is considered that the diffusion of the fragrant molecules in the gas phase occurs in the axial direction only (z, 1D). It is also assumed that the surrounding air is not soluble in the liquid, while simultaneous convection processes and interactions between the fragrant gas molecules and the surrounding air molecules are not occurring (ideal gas phase). In order to account for the transient state of the phases involved in this phenomenon, a mass balance is calculated over the gas control volume, $\Delta V = A\Delta z$, being V the total gas control volume, A the cross-section area, and z the axial coordinate, as shown in Fig. 3.4. There we have applied the Fick's Second Law for diffusion to describe the perfume diffusion model in the gas phase. Note, however, that Fick's Law does not account for interactions between gaseous molecules (and so is only suitable for a diluted gas, while the Maxwell–Stefan theory for diffusion could be used instead to take into account binary interactions for a concentrated gas phase) (Crank, 1975; Taylor and Krishna, 1993; Bird et al., 2002). This assumption seems reasonable for most perfume applications. In this way, it is possible to mathematically describe this process by the following partial differential equation (PDE):

$$\frac{\partial y_i}{\partial t} = D_{i,air}\frac{\partial^2 y_i}{\partial z^2} \tag{3.1}$$

where y_i represents the mole fraction of odorant i in the gas phase and is linked to the vapor concentration by the relationship: $y_i = c_i/c_T$, and $c_T = P/RT$ a constant value once the gas is considered ideal. The diffusion coefficient of odorant i is given by $D_{i,\text{air}}$ and t is time. Diffusion coefficients can be experimentally measured or calculated by correlations available in the literature. We have been using the method of Fuller et al. (for further details, see (Teixeira et al., 2009a)) which has proven to yield the smallest average error (Poling et al., 2004).

However, note, that the approach we are following is very similar to the well known Stefan tube problem (Lee and Wilke, 1954; Heinzelmann et al., 1965) which considers a liquid that is evaporating into a stagnant gas film with a stream of air sweeping at the top of the tube. The solution proposed by Stefan can be derived from Fick's First Law as a function of the molar flux relative to stationary coordinates as expressed by Bird et al. (1960). The slight difference between this approach and the one we are following here is that the former considers not only the diffusion molar flux of the fragrant species (diffusing with the current) but also its molar flux resulting from the bulk motion of the fluid (induced flow of the vapor species). Nevertheless, our studies have not shown significant differences between the two models for the specific problem studied here because we are dealing with highly diluted vapor concentrations (y_i is very small).

3.4.1.2 Liquid Phase

As previously discussed in this chapter, the liquid phase was considered as a nonideal mixture of fragrance ingredients dissolved in ethanol and water. The vapor–liquid equilibrium (VLE) assumed to be established at the liquid–gas interface was predicted using the UNIFAC method. Since the evaporation and diffusion processes of a mixture of fragrance ingredients is a transient process, a mass balance was defined in the liquid phase as presented by the following ordinary differential equation (ODE):

$$\frac{dn_i}{dt} = D_{i,air} A_{lg} c_T \frac{\partial y_i}{\partial z}\bigg|_{z=0} \tag{3.2}$$

where n_i is the number of moles of component i in the liquid phase and A_{lg} is the area of the liquid–gas interface. Once we are dealing with PDE, it is necessary to define initial and boundary conditions for our system.

Initial Conditions (IC)

Gas Phase

$$t = 0: \quad y_i = y_{i_0} = 0 \tag{3.3}$$

Liquid Phase

$$t = 0: \quad n_i = n_{i_0} \text{ or } x_i = x_{i_0} \tag{3.4}$$

where x_i is the mole fraction of component i in the solution and x_{i_0} is the initial mole fraction of component i in the liquid mixture.

Boundary Conditions (BC)

Two boundary conditions are defined ahead, at the left and at the right side of the domain of our system, as follows:

$t > 0$:

$$z = 0: y_i = \frac{\gamma_i P_i^{\text{sat}}}{P} x_i = \frac{\gamma_i P_i^{\text{sat}}}{P} \frac{n_i}{\sum_i n_i} \tag{3.5}$$

$$z = z_{\text{max}}: y_i = 0 \tag{3.6}$$

It is to be highlighted that the mole fraction of the ith component of a mixture of N components (y_i) at $z = 0$ represents the composition in equilibrium with the corresponding component mole fraction in the liquid phase (x_i). The VLE condition assumed at the interface was calculated using the UNIFAC method for the prediction of the activity coefficients (γ_i). It should be highlighted that our group has evaluated different group-contribution methods for estimating this parameter. The UNIFAC method has shown to be the best when compared with experimental data and proved to be perfectly suitable for accurately predicting the vapor compositions which allowed the calculation of odor intensities that were very close to olfactory evaluations performed by panelists. For further details on this topic, we encourage the reader to follow the literature (Teixeira et al., 2011). The numerical solution of the systems of PDE (ternary or quaternary mixtures will have six or eight paired differential equations, respectively) was computed in MATLAB using the *pdepe* package for numerical computation of PDE. The domain of integration and solution was defined by $z \in [z_{A_{gl}}, z_{\text{max}}]$, where $z_{A_{gl}} = 0$ corresponds to the liquid–gas interface. Furthermore, the time integration was calculated over the domain of

$t \in [0, t_{max}]$, where the limit on the right side could vary from hours to days, depending on the scope of the study. Further details on this model can be found in the literature (Crank, 1975; Cussler, 2007; Teixeira et al., 2009a).

3.4.2 Performance of Quaternary Fragrance Mixtures

The performance of several fragrance mixtures will be evaluated for both quaternary and quinary mixtures of the type top note + middle note + base note + solvent(s). The selected fragrance ingredients, their physicochemical properties, and the diffusion coefficients in air are presented in Table 3.3 while the mixtures are presented ahead:

1. limonene + geraniol + vanillin + ethanol,
2. limonene + geraniol + galaxolide + ethanol,
3. limonene + geraniol + vanillin + ethanol + water,
4. limonene + geraniol + galaxolide + ethanol + water.

The evaluation of the performance of these fragrances will address their release and propagation over time and distance from the source. First, the diffusion model was applied to simulate the evaporation and diffusion of the two quaternary mixtures with a selected initial composition as defined in Table 3.4. The obtained odor profiles for the simulations with mixture 1 are shown in Fig. 3.5, using both the

	Name	Molecular Formula	M_i (g/mol)	P_i^{sat} (Pa)	ODT_i (g/m³)	$\dfrac{P_i^{sat} \cdot M_i}{ODT_i \cdot RT}$	$D_{i,air}$ (m²/h)	n_i^d
A	Limonene[a]	$C_{10}H_{16}$	136.2	20.5×10^1	2.45×10^{-3}	4.60×10^3	2.214×10^{-2}	0.37
B	Geraniol[a]	$C_{10}H_{18}O$	154.3	26.7×10^{-1}	2.48×10^{-5}	6.70×10^3	2.138×10^{-2}	0.36
C	Vanillin[a]	$C_8H_8O_3$	152.2	16.0×10^{-3}	1.87×10^{-7}	5.25×10^3	4.111×10^{-2}	0.31
	Galaxolide	$C_{18}H_{26}O$	258.4	72.7×10^{-3b}	6.30×10^{-7c}	1.20×10^4	1.924×10^{-2}	0.36
S	Ethanol[a]	C_2H_6O	46.0	72.7×10^2	5.53×10^{-2}	2.44×10^3	4.469×10^{-2}	0.58
	Water	H_2O	18.01	3.17×10^3	–	–	9.132×10^{-2}	–

Table 3.3. Physicochemical properties for the selected fragrance chemicals

Diffusion coefficients calculated by Fuller et al. (1966). The ratio of odorant properties is a measure of the potency or blooming of a fragrance ingredient that can be used as a single component or within a mixture of fragrances.
[a]*From Calkin and Jellinek (1994).*
[b]*From Balk and Ford (1999).*
[c]*From Fráter et al. (1999).*
[d]*From Devos et al. (2002).*
Source: *Adapted with permission from Teixeira et al. (2009a). © 2009, Elsevier.*

Table 3.4. Comparison of the OV and Power Law Models at Different Times During Evaporation, for Perfumery Systems 1 and 2

		Mole Fractions				OV				Power Law			
	Time (s)	x_A	x_B	x_C	x_S	OV_A	OV_B	OV_C	OV_S	ψ_A	ψ_B	ψ_C	ψ_S
System 1—Mixture P1													
P1$_{initial}$	0.0	0.120	0.120	0.060	0.700	2947	840	343	1862	19	11	6	79
P1$_A$	14.4	0.288	0.391	0.196	0.125	3246	2396	3237	365	20	16	12	31
P1$_B$	72.1	0.192	0.497	0.250	0.061	2476	3108	3614	156	18	18	13	19
P1$_C$	198.4	0.110	0.565	0.286	0.039	1612	3653	3646	94	15	19	13	14
System 2—Mixture P1													
P1$_{initial}$	0.0	0.120	0.120	0.060	0.700	2142	764	2559	2038	17	11	17	83
P1$_A$	14.4	0.332	0.401	0.201	0.067	2089	3337	3421	361	17	19	19	30
P1$_B$	72.1	0.274	0.463	0.233	0.030	1730	3837	4032	156	16	19	20	19
P1$_C$	198.4	0.207	0.514	0.260	0.019	1340	4155	4640	94	14	20	21	14

Boxes highlight the component with the maximum odor intensity (dominant odor).
Source: *Adapted with permission from Teixeira et al. (2009a). © 2009, Elsevier.*

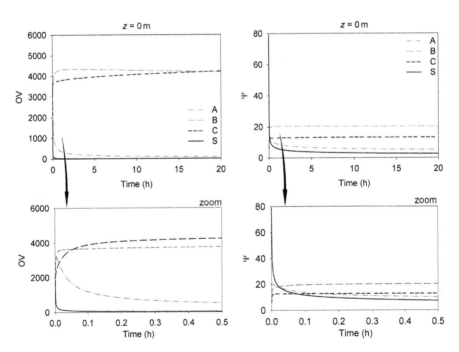

Fig. 3.5. Comparison between the odor profiles for mixture 1 using the OV (left) and the Power Law (right) as odor intensity models. These profiles show the evolution of the perceived odor near the point of release (z = 0 m). A: limonene, B: geraniol, C: vanillin, S: ethanol. Adapted with permission from Teixeira et al. (2009a). © 2009, Elsevier.

OV and Power Law models, for the initial mixture composition of 1 (see Table 3.4).

These odor profiles show some differences, starting by the different scales of the two models for describing odor intensity. However, apart from that it is seen the initial dominant odor is not the same if we use the OV or the Power Law: in the former case, it is limonene that is more strongly perceived, while in the last it is ethanol. Yet, despite that difference in the first odor impression (which corresponds to the performance parameter—impact—and is only perceived near the liquid—gas interface where we assumed VLE conditions), after some time of release the dominant odorant is geraniol for both cases. Such fact is clearly seen in Table 3.4: while it takes nearly 80 s for geraniol to become dominant when the Power Law model is used, around 200 s are needed with the OV model. Consequently, this is clearly not a significant difference in terms of odor perception, even more because when perfumers or customers are evaluating fragrances there are convection phenomena occurring simultaneously which speed up the release mechanism. Then, both odor intensity models give similar odor profiles over time. Note, however, that the high impact of ethanol predicted with the Power Law can be decreased by introducing water in the solution (as will be shown later) or using a polar fixative ingredient that will tend to retain in the liquid the more polar components (as it is often done in perfume formulation).

At this point, if we combine these fragrance diffusion data with the PTD® methodology presented in Chapter 2, it is possible to confirm the evolution of the perceived odor over time as shown in Fig. 3.6 for some selected times presented in Table 3.4. In the PTDs it is also represented the evaporation lines for the initial mixtures ($Pl_{initial}$) which show the evolution of the dominant scent over time (as well as the liquid composition). It is clear that these evaporation lines cross different odor zones as the evaporation takes place, resulting in a different olfactory perception over time. These PTDs simulated at different times (14.4 s, 72.1 s, and 198.4 s) during the release process show significant differences in the odor zones over time. As we have seen before, this is mostly due to the presence of the complex molecule of vanillin and the high exponent of ethanol for the Power Law. Vanillin interactions with the remaining ingredients are extremely difficult to predict because it is a highly multifunctional molecule (hydroxyl,

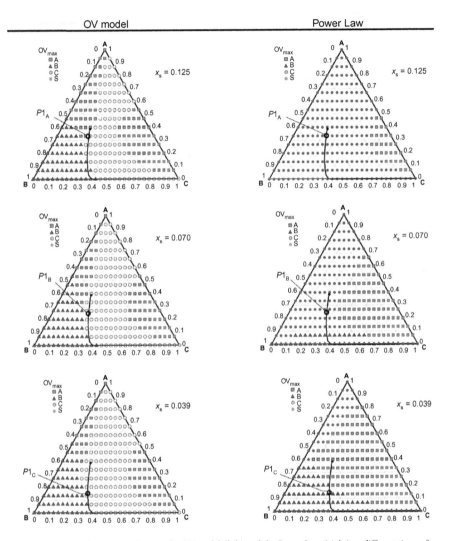

Fig. 3.6. PTD of perfume system 1, using the OV model (left) and the Power Law (right) at different times after application (14.4, 72.1, and 198.4 s). This perfumery system is composed of limonene (A) + geraniol (B) + vanillin (C) + ethanol (S). Adapted with permission from Teixeira (2011). © 2011, M.A. Teixeira.

ketone, and ether groups attached to an aromatic ring). In the case of the Power Law, it is seen that the ethanol odor intensity becomes significantly higher than any of the remaining fragrance ingredients and its odor zone dominates a large part of the PTD® as well. Nevertheless, the PTD® represents only the perception near the vapor–liquid interface or the source of release, and that is when the evaporation of ethanol is more significant. After the initial times, it is seen in the PTDs the same behavior

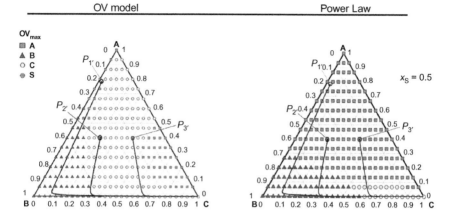

Fig. 3.7. Representation of the evaporation lines for two ternary mixtures with compositions $P_{1'} = [x_A = 0.780,$ $x_B = 0.200,$ $x_C = 0.200]$ and $P_{2'} = [x_A = 0.400,$ $x_B = 0.400,$ $x_C = 0.200],$ and a quaternary mixture of $P_{3''} = [x_A = 0.200,$ $x_B = 0.100,$ $x_C = 0.200,$ $x_S = 0.500],$ using the OV and Power Law models.

as that predicted by the diffusion model: geraniol becomes the dominant note either using the OV or the Power Law for intensity of odors.

It is also possible to observe how the perfume mixture evolves throughout the evaporation process for different initial compositions as presented in Fig. 3.7. Depending on the selected (initial) composition, evaporation lines cross different odor zones at distinct times, thus showing that the perceived odor will be different over time.

A similar analysis can be performed for mixture 2 where the base note has been changed to galaxolide. The corresponding diffusion profiles for this case are presented in Fig. 3.8 for the OV and Power Law models as well. In this case, the similarities of the odor profiles are visible for times over three min, as previously seen in Table 3.4 for the dominant odor (galaxolide dominates the odor character, followed by the middle note—geraniol—and top note—limonene—after this initial period of time). Once more, in the beginning of the release of the fragrance mixture there are some differences in the prediction of the odor character between the two models: in analogy with mixture 1, here it is seen that ethanol is also strongly perceived and dominates the first impact when the Power Law is used, while galaxolide is more strongly perceived when the OV model is applied.

As we know, during the release of the perfume mixture, odorant molecules evaporate and diffuse into the surrounding air above the

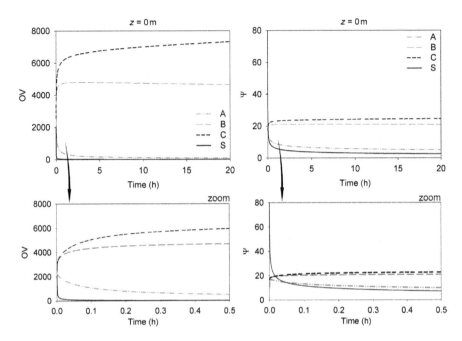

Fig. 3.8. Comparison of the odor profiles for perfume mixture P1 of the quaternary system 2, using the OV (left) and the Power Law (right) as odor intensity models. Adapted with permission from Teixeira et al. (2009a)). © 2009, Elsevier.

liquid. Throughout this slow process, the liquid composition is changing over time because fragrance molecules have different volatilities and interactions with the medium (thus different rates of evaporation and diffusion as well), and so the perceived odor character evolves with time and distance. Consequently, it is not a surprise for the reader that changing the initial composition of the perfume mixture may completely change the perceived odor (over time and distance): such fact means that it will be also changing the performance of the perfume.

Taking into account this effect between the liquid and the vapor phases, it is possible to represent evaporation lines in the PQ2D® to show the evolution of the odor and predict its future perception. Fig. 3.9 represents evaporation lines for the two quaternary mixtures studied here. These lines represent the paths of the perfume mixtures (the evolution of the composition in the remaining perfume) as they are evaporating and diffusing in the air. The color scheme allows identifying the dominant odorants considering the Power Law model. It is seen that throughout the release and diffusion of the perfume mixtures, evaporation lines cross different odor volumes in the PQ2D®

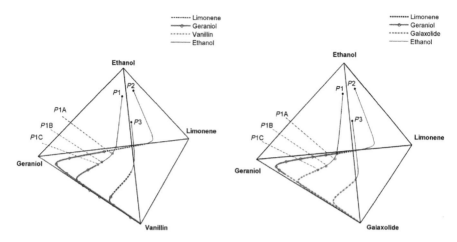

Fig. 3.9. Perfumery evaporation lines plotted in the PQ2D® for the quaternary system 1 (left) and system 2 (right) with different initial mixture compositions. Adapted with permission from Teixeira et al. (2010). © 2010, John Wiley & Sons.

(the color of the evaporation line changes over time) which means that the dominant smell is changing. Note, however, that the shape of the evaporation path is strongly dependent on the initial mixture composition, though they all follow the same trend: initially, they are characterized by a steep curve due to the fast evaporation of ethanol. Then the evaporation lines approximate the geraniol–vanillin or geraniol–galaxolide Perfumery Binary Surface, because limonene is very volatile. Finally, evaporation lines tend to approach the vertex of the base note (vanillin or galaxolide) since it is the heaviest fragrance component.

Finally, the evaluation of the performance of fragrance mixtures can be better visualized if we represent the dominant odor in a 2D plot of distance versus time, as shown in Fig. 3.10. This representation is called a performance plot in analogy with Fig. 3.1 where we can clearly see where the four performance parameters fit in the distance versus time diagram. For this particular quaternary mixture with the selected composition of Table 3.4 we see that ethanol dominates the initial impact of the perfume mixture, just followed by limonene. The other performance parameters are dominated by geraniol for tenacity and vanillin for diffusion and volume properties. Such behavior for a simple perfume mixture shows how the perceived odor may change with time and, consequently, how is the evolution of the performance of the fragrance. It should not be neglected that although

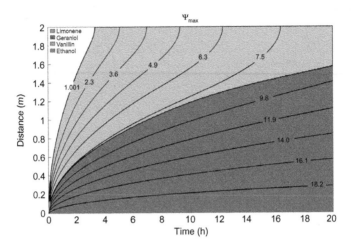

Fig. 3.10. *Performance plot for the mixture limonene + geraniol + vanillin + ethanol showing the isolines for odor intensity.* Adapted with permission from Teixeira (2011). © 2011, M.A. Teixeira.

we are performing this analysis in respect to the dominant odor, the perceived mixture will always be a blend of the different fragrance ingredients involved.

3.4.3 Performance of Quinary Fragrance Mixtures

The majority of the fragrances available in the market possess water in the formulation or are designed to be applied in aqueous systems (e.g., detergents or shampoos). Moreover, water is a very unique molecule which has some particularities like high polarity and strong ability to establish strong hydrogen bonding. Consequently, the effect of water might be of relevance in terms of fragrance release and propagation in air. We have previously addressed partially this topic in Chapter 2 when we evaluated the effect of water in the impact of the initial odor. To evaluate the effect of water on perfume performance, we can just add water to the perfume mixtures used before (mixtures 1 and 2 in Table 3.4). These new mixtures, called 3 and 4, and presented in Table 3.5, are similar in terms of composition to the previous mixtures 1 and 2 (number of moles in the initial quinary mixture is equal to the quaternary ones, for all the fragrance ingredients and ethanol, thus only water was added in a mole fraction of 0.45). Consequently, the composition of quinary systems in a water free basis is the same as the composition of the corresponding quaternary systems. The evaporation profiles of the two quinary perfume mixtures are presented in Fig. 3.11 using the Power Law as the odor intensity

Table 3.5. Odor Intensity Values Using the Power Law Model at Different Times During Evaporation for the Quinary Mixtures 3 and 4

		Mole Fractions					Power Law				
	Time (s)	x_A	x_B	x_C	x_S	x_{H_2O}	ψ_A	ψ_B	ψ_C	ψ_S	ψ_{H_2O}
System 3—Mixture P1											
$Pl_{initial}$	0.0	0.066	0.066	0.033	0.385	0.450	24	11	3	53	–
Pl_A	14.4	0.233	0.331	0.166	0.145	0.125	20	16	10	32	–
Pl_B	72.1	0.173	0.466	0.234	0.063	0.064	18	18	12	19	–
Pl_C	198.4	0.103	0.541	0.273	0.040	0.044	15	19	12	14	–
System 4—Mixture P1											
$Pl_{initial}$	0.0	0.066	0.066	0.033	0.385	0.450	21	10	16	54	–
Pl_A	14.4	0.301	0.368	0.184	0.078	0.069	17	18	18	32	–
Pl_B	72.1	0.262	0.448	0.225	0.032	0.033	16	19	20	19	–
Pl_C	198.4	0.201	0.503	0.254	0.020	0.023	14	20	21	14	–

Grey Boxes represent maximum odor intensity. Although water is a chemical component within this mixtures, it is not perceived by the human nose.
Source: *Adapted with permission from Teixeira et al. (2010).* © *2010, John Wiley & Sons.*

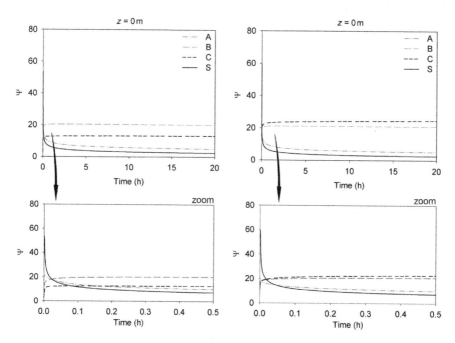

Fig. 3.11. *Evaporation profiles for the quinary mixtures of system 1 (left) and system 2 (right) using the Power Law model.*

model. Furthermore, the predicted compositions for some selected times after release has started are shown in Table 3.5.

If we compare the quaternary (Table 3.4) and the quinary mixtures (Table 3.5) for the Power Law model, it is possible to see that in the latter, ethanol is less dominant when water is introduced in the formulation (although there is also some dilution effect). This is a fact previously observed by other authors as we have reported in Chapter 2 when addressing the effect of matrix on the release of fragrances. Clearly, water has a retention effect on ethanol which can be explained by the strong affinity (hydrogen bonding) between them. Consequently, it slows down the evaporation of ethanol from the liquid mixture, thus reducing its high initial odor intensity. The presence of water in the perfume formulation provides another effect: it enhances the release of the top note, limonene. This can be explained by the fact that the higher the water content, the higher the polarity of the mixture will be and so it will tend to push out of the solution the nonpolar fragrance ingredients (like limonene).

Once more, although ethanol presents a high odor impact (in the first moments) when using the Power Law model, we are considering here only diffusion effects. If convection is considered, it would make the results from OV and Power law models similar. Furthermore, the fragrance mixtures studied here are simple with only three to five components, while commercial perfumes, for example, may have dozens of compounds in their formulation, some of which will play decisive roles on the release of others.

3.4.4 Evaluation of Commercial Perfumes

So far, we have been studying ternary to quinary mixtures of a defined and known composition. However, it is unquestionable that for the fragrance business it is common practice to use larger number of fragrance ingredients in a formulation. Perfume mixtures can differ in complexity and be applied in different products with different bases/matrices. Here, the evaporation and diffusion of two commercial perfumes will be addressed without delving into much detail due to the complexity of its nature and the difficulty of its validation. The odor profiles for Gloria (Cacharel) and L'air du Temps (Nina Ricci) are presented in Figs 3.12 and 3.13, respectively.

These commercial perfumes were analyzed by gas chromatography coupled with mass spectrometry (GC-FID-MS) for the quantification

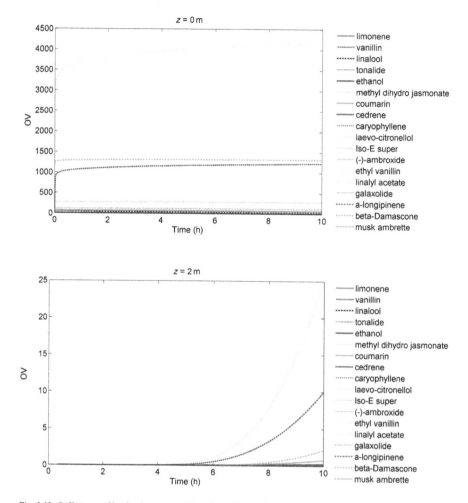

Fig. 3.12. Diffusion profiles for the commercial perfume Gloria (Cacharel) at z = 0 m and z = 2 m from the releasing source. Adapted with permission from Teixeira (2011). © 2011, M.A. Teixeira.

and identification of the major fragrance ingredients. Nevertheless, the purpose here is to show a qualitative perspective obtained for more complex mixtures (closer to real perfumes). From that point, the odor model presented previously in this chapter was applied for the diffusion of the different components and the OV concept was used due to the large number of components in the mixture (and lack of available Power Law exponents). It is possible to observe that there is a large number of fragrance ingredients in these perfumes with different roles in it. Some are more strongly perceived than others (higher OV) depending on the time and distance from the source, thus contributing

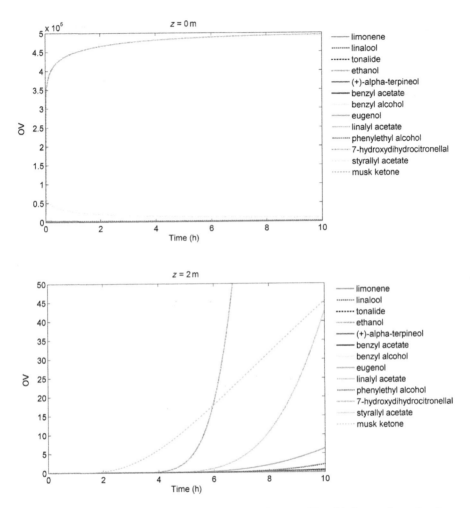

Fig. 3.13. *Diffusion profiles for the commercial perfume L'air du Temps (Nina Ricci) at* $z = 0\,m$ *and* $z = 2\,m$ *from the releasing source.* Adapted with permission from Teixeira (2011). © 2011, M.A. Teixeira.

for the perceived overall odor. This is one of the reasons behind the complexity and success of perfumes.

3.5 CONCLUSION

In this chapter, we have presented our approach to predict the performance of fragrance ingredients and mixtures. Four key performance parameters are those which account for the diffusion of the fragrance notes, that is, the effect of time and distance on the perception of the ingredients. These performance parameters are the so-called Impact,

Diffusion, Volume, and Tenacity. The first part of the chapter was devoted to the evaporation of the ingredients. Using models from Thermodynamics (group-contribution models), we are able to predict activity coefficients. These activity coefficients can then be used to predict the vapor compositions of the ingredients in equilibrium with the liquid perfume. We have demonstrated that a good accuracy can be obtained in such predictions, both in terms of compositions and in terms of predicted odor (when the model is combined with an odor intensity and perception model). Then a diffusion model, based on Fick's Second Law, was presented. This model, based on PDEs to be solved simultaneously, allows predicting the diffusion profiles of each ingredient in the perfume formulation. We have showed the results for several examples, from quaternary mixtures to some commercial perfume formulations. The diffusion profiles allow identifying the key ingredients responsible for the *impact, diffusion, volume,* and *tenacity* of the perfume, or in other words, the performance of fragrances. Moreover, a new type of performance diagram is presented that shows these four key performance parameters together in a 2D odor map. This "performance plot" (see Fig. 3.10) provides in a simple perspective the main olfactory effects of a perfume formulation in terms of time and distance.

REFERENCES

Abbe, N. J. V., Perfumes: their manufacture in products and psychology in use, *Poacher's Perfumes, Cosmetics and Soaps*, Butler, Dordrecht, The Netherlands, Kluwer Academic Publishers, 2000.

Aikens, P. A., Zhang, Z., Friberg, S. E., Change of amphiphilic association structures during evaporation from emulsions in surfactant-fragrance-water systems, *International Journal of Cosmetic Science*, 22: 181–199, 2000.

Al-Bawab, A., Odeh, F., Bozeya, A., Aikens, P., Friberg, S., A comparison between the experimental and estimated evaporation paths from emulsions, *Flavour and Fragrance Journal*, 2009.

Arce, A., Marchiaro, A., Rodriguez, O., Soto, A., Liquid–liquid equilibria of limonene plus linalool plus diethylene glycol system at different temperatures, *Chemical Engineering Journal*, 89 (1–3): 223–227, 2002.

Balk, F., Ford, R., Environmental risk assessment for the polycyclic musks AHTN and HHCB in the EU—I. Fate and exposure assessment, *Toxicology Letters*, 111: 57–79, 1999.

Baydar, A., Carles, A., Decazes, J. M., McGee, T., Purzycki, K., Behavior of fragrances on skin, *Cosmetics & Toiletries*, 111: 49–57, 1996.

Baydar, A., McGee, T., Purzycki, K. L., Skin odor value technology for fragrance performance optimization, *Perfumer & Flavorist*, 20: 45–53, 1995.

Behan, J. M., Macmaster, A. P., Perring, K. D., Tuck, K. M., Insight into how skin changes perfume, *International Journal of Cosmetic Science*, 18, (5): 237–246, 1996.

Behan, J. M., Perring, K. D., Perfume interactions with sodium dodecyl-sulfate solutions, *International Journal of Cosmetic Science*, 9 (6): 261−268, 1987.

Bird, R., Stewart, W., Lightfoot, E., Transport Phenomena, Singapore, John Wiley & Sons, 1960.

Bird, R., Stewart, W., Lightfoot, E., Transport Phenomena, New York, John Wiley & Sons, 2002.

Bozeya, A., Al-Bawab, A., Friberg, S. E., Guo, R., Equilibration in a geranyl acetate emulsion, *Colloids and Surfaces A: Physicochemical and Engineering Aspects*, 373 (1−3): 110−115, 2009.

Brossard, C., Rousseau, F., Llamas, G., Dumont, J. P., Determination of sensory oil-water, partition coefficients of single aroma compounds combining panelist free intensity rating and theoretical modeling of odor intensity, *Journal of Sensory Studies*, 17: 445−460, 2002.

Burr, C., The Perfect Scent: A Year Inside the Perfume Industry, Paris, Henry Holt and Co., 2008.

Calkin, R., Jellinek, S., Perfumery: Practice and Principles, New York, John Wiley & Sons, 1994.

Cortez-Pereira, C. S., Baby, A. R., Kaneko, T. M., Velasco, M. V. R., Sensory approach to measure fragrance intensity on the skin, *Journal of Sensory Studies*, 24 (6): 871−901, 2009.

Crank, J., The Mathematics of Diffusion, Oxford, Oxford Science Publications, 1975.

Cussler, E. L., Diffusion: Mass Transfer in Fluids, Cambridge, UK, Cambridge University Press, 2007.

de Doz, M. B. G., Cases, A. M., Solimo, H. N., Liquid plus liquid) equilibria of (water plus linalool plus limonene) ternary system at $T = (298.15, 308.15, \text{ and } 318.15)$ K, *Journal of Chemical Thermodynamics*, 40 (11): 1575−1579, 2008.

Devos, M., Rouault, J., Laffort, P., Standardized olfactory power law exponents, Dijon—France, Editions Universitaires-Sciences, 2002.

Domanska, U., Morawski, P., Piekarska, M., Solubility of perfumery and fragrance raw materials based on cyclohexane in 1-octanol under ambient and high pressures up to 900 MPa, *Journal of Chemical Thermodynamics*, 40 (4): 710−717, 2008.

Domanska, U., Paduszynski, K., Niszczota, Z. K., Solubility of fragrance raw materials in water experimental study, correlations, and Mod UNIFAC (Do) predictions, *Journal of Chemical Thermodynamics*, 43 (1): 28−33, 2010.

Duprey, R. J. H., Perring, K. D., Ness, J. N., Perfume Compositions − US Patent 7,713,922 B2. US, Quest International Services B.V.2010.

Escher, S. D., Oliveros, E., A Quantitative Study of Factors That Influence the Substantivity of Fragrance Chemicals on Laundered and Dried Fabrics, *Journal of the American Oil Chemists' Society*, 71, (1): 1994.

Fadel, A., Turk, R., Mudge, G., Mattila, J., Esteves, J., Ranciato, J., Perfumes for rinse-off systems. Corporation. United States Patent. US 7,446,079 B2, 2008.

Fadel, A., Turk, R., Mudge, G., Sullivan, D., Goberdhan, V., Meo, A. D., Malodor covering perfumery - US Patent 7,585,833 B2. Corporation. US2009.

Fráter, G., Müller, U., Kraft, P., Preparation and olfactory characterization of the enantiomerically pure isomers of the perfumery synthetic galaxolide, *Helvetica Chimica Acta*, 82: 1656−1665, 1999.

Friberg, S. E., Fragrance compounds and amphiphilic association structures, *Advances in Colloid and Interface Science*, 75, (3): 181−214, 1998.

Friberg, S. E., Evaporation from Three Single-Compound Phases in Equilibrium, *Journal of Dispersion Science and Technology and Health Care*, 28, (2): 207−212, 2007.

Friberg, S. E., Phase diagram approach to evaporation from emulsions with n oil compounds, *Journal of Physical Chemistry B*, 113 (12): 3894–3900, 2009.

Friberg, S. E., Al-Bawab, A., Odehb, F., Bozeya, A., Aikens, P. A., Emulsion evaporation path. A first comparison of experimental and calculated values, *Colloids and Surfaces A: Physicochemical and Engineering Aspects*, 338 (1–3): 102–106, 2009.

Friberg, S. E., Al-Bawab, A., Odehb, F., Bozeya, A., Aikens, P. A., Emulsion evaporation path. A first comparison of experimental and calculated values, *Colloids and Surfaces a-Physicochemical and Engineering Aspects*, 338, (1–3): 102–106, 2010.

Fuller, E. N., Schettler, P. D., Giddings, J. C., A new method for prediction of binary gas-phase diffusion coefficients, *Industrial and Engineering Chemistry*, 58, (5): 19–27, 1966.

Gomes, P. B., Engineering Perfumes. Department of Chemical Engineering. Porto, University of Porto Faculty of Engineering. PhD Thesis, 2005.

Guy, R. H., Predicting the rate and extent of fragrance chemical absorption into and through the skin, *Chemical Research in Toxicology*, 23 (5): 864–870, 2010.

Gygax, H., Koch, H., The Measurement of Odours, *Chimia, Special Issue in Flavours and Fragrances*, 55: 401–405, 2001.

Haefliger, O. P., Jeckelmann, N., Ouali, L., Leon, G., Real-Time Monitoring of Fragrance Release from Cotton Towels by Low Thermal Mass Gas Chromatography Using a Longitudinally Modulating Cryogenic System for Headspace Sampling and Injection, *Analytical Chemistry*, 82, (2): 729–737, 2010.

Heinzelmann, F. J., Wasan, D. T., Wilke, C. R., Concentration profiles in a Stefan diffusion tube, *Industrial & Engineering Chemistry Fundamentals*, 4 (1): 55–61, 1965.

Heltovics, G., Holland, L. A. M., Warwick, J. M., Jenkins, D. M., Sutton, K. L., Pretswell, E. L., Shefferd, A. J. P., Methods and compositions for improved fragrancing of a surface – US Patent 2004/0097398 A1. 2004/0097398 A1. Company. US. 2004/0097398 A12004.

Kasting, G. B., Saiyasombati, P., A physico-chemical properties based model for estimating evaporation and absorption rates of perfumes from skin, *International Journal of Cosmetic Science*, 23: 49–58, 2001.

Kasting, G. B., Saiyasombati, P., Two stage kinetics analysis of fragrance evaporation and absorption from skin, *International Journal of Cosmetic Science*, 25: 235–243, 2003.

Kasting, G. B., Saiyasombati, P., In vivo evaporation rate of benzyl alcohol from human skin, *Journal of Pharmaceutical Sciences*, 93, (2): 515–520, 2004.

Kerschner, J., Fragrances in Consumer Products, International Flavor and Fragrances—Bergen Academy, 2006.

Lee, C. Y., Wilke, C. R., Measurements of vapor diffusion coefficient, *Industrial and Engineering Chemistry*, 46 (11): 2381–2387, 1954.

Martel, B., Morcellet, M., Ruffin, D., Vinet, F., Weltrowski, M., Capture and Controlled Release of Fragrances by CD Finished Textiles, *Journal of Inclusion Phenomena and Macrocyclic Chemistry*, 44: 439–442, 2002.

Mata, V. G., Gomes, P. B., Rodrigues, A. E., Engineering Perfumes, *AICHE Journal*, 51, (10): 2834–2852, 2005.

Mookherjee, B., Patel, S., Trenkle, R., Wilson, R., A Novel Technology to Study the Emission of Fragrance from the Skin, *Perfumer&Flavorist*, 1, (23): 1998.

Poling, B., Prausnitz, J. M., O'Connell, J., The Properties of Gases and Liquids, McGraw-Hill, 2004.

Rodrigues, S. N., Martins, I. M., Fernandes, I. P., Gomes, P. B., Mata, V. G., Barreiro, M. F., Rodrigues, A. E., Scentfashion®: Microencapsulated perfumes for textile application, *Chemical Engineering Journal*, 149: 463–472, 2009.

Ryan, D., Prenzler, P. D., Saliba, A. J., Scollary, G. R., The significance of low impact odorants in global odour perception, *Trends in Food Science & Technology*, 19 (7): 383–389, 2008.

Schwarzenbach, R., Bertschi, L., Models to assess perfume diffusion from skin, *International Journal of Cosmetic Science*, 23, (2): 85–98, 2001.

Spedding, P. L., Grimshaw, J., O'Hare, K. D., Abnormal Evaporation Rate of Ethanol from Low Concentration Aqueous Solutions, *Langmuir*, 9: 1408–1413, 1993.

Specos, M. M. M., Escobar, G., Marino, P., Puggia, C., Victoria, M., Tesoriero, D., Hermida, L., Aroma Finishing of Cotton Fabrics by Means of Microencapsulation Techniques, *Journal of Industrial Textiles*, 40, (1): 13–32, 2010.

Stora, T., Eschera, S., Morris, A., The Physicochemical Basis of Perfume Performance in Consumer Products, *Chimia*, 55: 406–412, 2001.

Taylor, R., Krishna, R., Multicomponent Mass Transfer, New York, John Wiley & Sons, 1993.

Teixeira, M. A., Perfume Performance and Classification: Perfumery Quaternary-Quinary Diagram (PQ2D$^®$) and Perfumery Radar. Department of Chemical Engineering, Faculty of Engineering of University of Porto, PhD Thesis, 2011.

Teixeira, M. A., Rodríguez, O., Mata, V. G., Rodrigues, A. E., The diffusion of perfume mixtures and odor performance, *Chemical Engineering Science*, 64: 2570–2589, 2009a.

Teixeira, M. A., Rodríguez, O., Mata, V. G., Rodrigues, A. E., Perfumery quaternary diagrams for engineering perfumes, *AIChE Journal*, 55 (8): 2171–2185, 2009b.

Teixeira, M. A., Rodríguez, O., Rodrigues, A. E., The perception of fragrance mixtures: A comparison of odor intensity models, *AIChE Journal*, 56 (4): 1090–1106, 2010.

Teixeira, M. A., Rodríguez, O., Mota, F. L., Macedo, E. A., Rodrigues, A. E., Evaluation of group-contribution methods to predict VLE and odor intensity of fragrances, *Industrial & Engineering Chemistry Research*, 50: 9390–9402, 2011.

Teixeira, M. A., Rodríguez, O., Rodrigues, S., Martins, I., Rodrigues, A. E., A case study of Product Engineering: Performance of microencapsulated perfumes on textile applications, *AIChE Journal*, 58, (6): 1939–1950, 2011b.

Ternat, C., Ouali, L., Sommer, H., Fieber, W., Velazco, M. I., Plummer, C. J. G., Kreutzer, G., Klok, H. A., Manson, J. A. E., Herrmann, A., Investigation of the Release of Bioactive Volatiles from Amphiphilic Multiarm Star-Block Copolymers by Thermogravimetry and Dynamic Headspace Analysis, *Macromolecules*, 41, (19): 7079–7089, 2008.

Tokuoka, Y., Uchiyama, H., Abe, M., Phase-diagrams of surfactant water synthetic perfume ternary-systems, *Colloid and Polymer Science*, 272 (3): 317–323, 1994.

Vuilleumier, C., Flament, I., Sauvegrain, P., Headspace analysis study of evaporation rate of perfume ingredients applied onto skin, *International Journal of Cosmetic Science*, 17: 61–76, 1995.

Classification of Perfumes—Perfumery Radar

The qualitative classification of fragrances by words or descriptors is a difficult task, and so its prediction (using any kind of theoretical model) is even more complex. If the former is addressed by experts in perfumery, the latter is mostly considered as utopian, at least in forthcoming years. That is why F&F companies have been doing this job on the basis of sensorial analysis performed by their experienced perfumers, who use multiple (and sometimes complex) odor descriptors to map the olfactory space. However, because the universe is not governed by laws that approach to ideality but instead foresee the increasing of entropy, the way we perceive odor quality varies from person to person and so remains difficult to explain. Consequently, there are significant differences between the classifications obtained from different F&F companies. As a result, once again, human empiric knowledge and expertise rules this job. Altogether, these facts make it virtually impossible to believe in a classification of fragrances with universal acceptance.

Still, wouldn't it be wonderful to establish a systematic classification of fragrances using scientific knowledge, and so reducing interpersonal variability? The avid search for the answer to this question was probably the biggest motivation for the development of what would be named, the perfumery radar (PR). This was a long-developed idea which started in 2005 at the LSRE, and was first presented to the scientific community in 2004 (Mata et al., 2004) and later published in an international scientific journal in 2010 (Teixeira et al., 2010). It turned out to be a highly appreciated work with various endorsements and advertisings in both national and international press. Surprisingly or not, it got attention from the international community long before the national one. But before going into the details of this novel tool for perfume classification, let us explore further the way we classify odors, highlighting also the obstacles we find doing it.

4.1 THE PERCEPTION OF ODORS

The classification of odors with descriptors has long been studied though we were not able yet to fully understand the process: interpreting the biochemical processes behind olfaction is akin to looking for a needle in a haystack. Consequently, there is not a universal classification of odors into olfactory classes (Chastrette, 1998; Distel et al., 1999; Lawless, 1999; Pintore et al., 2006; Gilbert, 2008; Haddad et al., 2008; Zarzo and Stanton, 2009). The F&F industry should be understood as an R&D-oriented business, and these departments receive huge investments: construction of equipments and development of techniques for extraction of essential oils and chemical analysis, or finding new synthetic routes for fragrant molecules. Despite this, they are still far from mastering other topics such as the simulation of odor quality. The reason behind this lack of knowledge is due to the fact that odor recognition involves neuronal decoding, apart from the physiological and biochemical processes occurring at the nose level. The human nose, together with the olfactory receptor (OR) cells and their transduction within the brain, results in a complex but somewhat limited system for perceiving odors. In fact, we may be able to detect odorants at lower concentrations than a gas chromatographic equipment but, at the same time, we can only distinguish few chemicals presented in a mixture of hundreds of different odors (e.g., coffee has nearly 800 chemical compounds in its composition: How many can a consumer perceive?

And an expert? As we will see, this number will certainly be small; Parliment, 2002).

Apart from the great progresses made in the recent past on olfaction, we are still taking small steps in understanding all the processes within its mechanism. The relevant work of Buck and Axel on the discovery of the large family of ORs revolutionized the understanding of olfaction (Buck and Axel, 1991; Axel, 2005; Buck, 2005). It was not surprising, they were awarded the 2004 Nobel Prize in Physiology or Medicine. Additionally, in food chemistry and flavor science much has been done aiming to characterize the odor of food and drinks (Fisher and Scott, 1997). For example, different studies were performed to evaluate the aroma profile of wines, spirits, and soft drinks, often using radar plots to relate the composition with their odor character based on perceptual judgments of enologists or flavorists (Ferreira et al., 2000; Buettner and Schieberle, 2001; Caldeira et al., 2002; Moyano et al., 2002). So a relevant question is legitimate to ask: why is the olfactory system so unknown when compared to other senses?

Whatever is the physiology behind that, the sense of smell is keener than that of any other sense. A practical example to illustrate this fact is to compare colors and smells: usually, a general description of a color is done through the use of one or two words (at most) which often explicitly translates the color that any other individual with good visual acuity would also describe. On the opposite, the same is often not true for the description of odors. When one tries to describe the smell elicited by a rose, for example, it will require multiple words, representing feelings, emotions, or memories of the individual. Moreover, different people will use different adjectives or classes to characterize the scent of a rose. The main reason for that is because we are not taught for different odors, as we are for colors, shapes, or sounds. Other reasons are related to the interpersonal variability and the complexity of the human olfactory system. Once again we are back to the question of how the human being interprets odorant molecules and characterizes them in terms of odors. For the human vision, we know that there are three broadly tuned receptors that perceive the entire visible range of wavelengths. However, in olfaction we have about 391 olfactive receptors that allow us to perceive thousands of different odorants and multiple combinations between them (Saito et al., 2009).

Several examples can be found in nature where enigmatic effects on odor perception are observed. In some cases, mixtures of odors can produce unpredictable odor perceptions. Even among single odorants, some are perceived differently as their concentration increases. This means that an odorant is not perceived with the same quality at low and high concentrations. The general trend is that unpleasant odors are perceived at high concentrations. For example, some thiols evidence a harmonious fruity odor at very low concentrations (near its odor detection threshold) but unpleasant (and sometimes nauseous) sulfurous odors at higher concentrations. One other puzzling odorant molecule is trans-non-2-enal, because its perceived character is highly appreciated for fragrance purposes at low concentrations but is unpleasant for high ones (Fisher and Scott, 1997). Since its odor quality changes with the concentration, it has different recognition thresholds, and that is why it appears to be classified as a green, cucumber, aldehydic, and fatty odorant with a citrus *nuance*.[1] Other examples were observed for macrocyclic ketones, which have a faint cedarwood odor when concentrated and a musky odor when diluted (Chastrette, 1998).

However, within the F&F industry the task of sensory perception, evaluation, and classification of odors or mixtures is part of the job of perfumers. These specialists are highly trained for many years to detect, recognize, and classify raw materials and complex perfumes. The nose of a perfumer is capable of recognizing a few hundreds of different odorant chemicals, keeping them as a memory database. Their classification generally consists of assigning odors to olfactory families or classes, which often include *nuances*. The number of classifications available (either from top F&F companies or experts) is large, but the agreement between them is limited. Typical olfactory families are citrus, floral, green, fruity, herbaceous, musk, oriental, spicy, tobacco, woody, chypre, aromatic, or fougère, among many others, although they often change within fragrance houses or perfumers. These terms can be seen as descriptors of the type of odor, and are commonly sided by quality terms like heavy or light, sharp or round, just to mention a few. It is important to highlight that this nomenclature is difficult to be interpreted by the typical consumer, who is not familiarized with the terminology. Nonexperts usually classify odors using more traditional terms and expressions that resemble some of

---

[1]A nuance is a subtle or secondary odor in the main olfactory sensation of a pure chemical or a perfume.

their memories or past experiences. However, even among perfumers there is not a complete agreement of which olfactory families should be used or in which to include the perceived scents. That is, there is not (still) a glossary widely accepted for olfactive descriptors, despite some efforts done in that direction (Ellena, 1987; Chastrette, 1998; Zarzo, 2008; Zarzo and Stanton, 2009). Yet, there are several reasons that contribute to this uncertainty in olfaction (Milotic, 2003):

1. Perception of fragrances presents interpersonal variability (the human olfactory system presents sensory-chemical differences, e.g., physiological).
2. It is very difficult to characterize fragrances and scents by words, since people are not taught for different odors as they have been for colors, shapes, or sounds. Perfumers are an exception, of course.
3. People tend to associate smells to past experiences, objects, feelings, and emotions, which lead to a variety of classifications. Perfumers can be considered an exception: they tend to use the same language or olfactory descriptors, which is more accurate (Jaubert et al., 1995).
4. Science behind perfume formulation is proprietary, but the success of commercial perfumes is always unpredictable because it is mainly consumer-driven. And there is no secret formula to predict that.

Thus, if we attempt to put into words how we perceive odors and what influences us when recognizing and classifying them, the proposed pyramid of scents shown in Fig. 4.1 might give a simple perspective of the hierarchical processes within the perception of scents (Teixeira et al., 2010). Briefly, it includes four layers, cross-referenced, which a perfumer must have but a nonexpert will hardly achieve. At the bottom of the pyramid we have our emotions, which are based on our culture, memories, and past experiences (Distel et al., 1999;

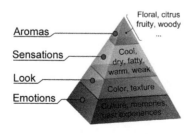

Fig. 4.1. The pyramid of scents perception. Adapted with permission from Teixeira et al. (2010). © 2010, American Chemical Society.

Wilson, 2009). It is this primary knowledge that a person (especially the nontrained) will start using to characterize a perceived fragrance. From that, his or her judgment will also be influenced by other sensory stimuli (sight, touch, taste, and hearing) which we defined in Fig. 4.1 as the look. It is only after these two layers that we really start the classification of a fragrance: on the third layer with a subjective analysis, and subsequently with an objective classification. In the former, a large number of different sensations expressed by words like cool or warm, dry, fatty or powerful, and so on, are commonly used to describe fragrances, especially by nontrained people (Milotic, 2003). In fact, when naïve subjects are asked to classify odorants they are strongly influenced by emotional, subjective, and past experiences (Chastrette, 2002). This classification is often ambiguous, individual dependent, and difficult to materialize. And this is the point where a nontrained person will generally be limited for the classification of odors. This is so, because at the top of the pyramid of scents relies the most objective classification, which is often given by perfumers (Milotic, 2003). It uses a more formal terminology which is not within the reach of nontrained people (regular consumers). However, even these classification terms may also be ambiguous and different for the same fragrance ingredient or perfume mixture.

The interpretation and discrimination of odors seem to be easier for pure chemicals than for complex mixtures (Gilbert, 2008). Even though, the attribution of a class or family to certain mixtures of scents might become simpler than for single chemicals (e.g., coffee blends or essential oils), the fact is that in the former what is being identified by the nose is only a limited number of components while the background ones are being neglected. This idea was first tackled by Laing (Laing, 1983, 1987; Gilbert, 2008), intrigued by the number of smells that the nose alone could pick out from a complex mixture. In his work, Laing showed that neither nonexperts nor experts could identify more than three to four odors (or volatile components) from mixtures. Moreover, as further odors are added to the mixture, the difficulty to identify even one of them increased. Recalling that the human nose can detect single smells at extraordinarily low concentrations, Gilbert concluded that *we do a better job of collecting smells than we do of tracking them in a complex mixture* (Gilbert, 2008). In brief, the classification of single chemicals might be more concise, complete, and truthful of its whole character than it is for mixtures (Jinks and Laing, 2001; Gilbert, 2008).

In the specific case of perfumes, they can be classified in terms of the concentration of the perfume concentrate (e.g., *Eau de Parfum, Eau de Toillete, Eau Frâiche*) or in terms of its olfactory family (e.g., floral, citrus, and woody) (Pybus and Sell, 1999; Zarzo, 2008; Zarzo and Stanton, 2009). The amount of concentrate or essential oil in a perfume formulation can vary widely, depending on the purpose of the perfumed product. The composition of fragrance components ranges from 10% to 30% for some compounds, down to trace levels (parts per million) for others. Solvents are also used in the formulation process, ethanol and water being the most common, as well as diethyl phthalate (DEP) or dipropylene glycol (DPG) (Schreiber, 2005; Sell, 2006; Surburg and Panten, 2006).

Moreover, just as there are classifications of fragrances into olfactory families, there are also other organoleptic classifications for different types of products that are strongly dependent on sensorial perception. Some examples are the olfactory classification of wines and beers as shown in Figs 4.2 and 4.3, respectively. The Beer Wheel presented in Fig. 4.3 dates back to the 1970s and considers both the classifications of flavors (gustatory sense) and odors (olfactory sense) of beers. It was developed by Morten Meilgaard and was later adopted as the flavor analysis standard by the European Brewery Convention, the American Society of Brewing Chemists, and the Master Brewers Association of the Americas (Bamforth et al., 2008).

4.1.1 Classification of Odors

As previously discussed, the classification of odors is an extremely difficult job that at the F&F industry is deliberately given to experts, the perfumers. For those with a less discerning nose, such task is almost impossible to be performed, while finding subtle differences between odors are merely utopian. Due to this reason, we will start to discuss different approaches developed over the years for the classifications of single odorants and will only later address their mixtures.

The majority of these classifications were based on the olfactory evaluation of such experts which the industry still follows today. Nevertheless, over the past two centuries, different scientists from diverse scientific fields around the globe have searched for other means of describing the human olfactory space based on (i) empirical classifications, (ii) stereochemical theories and primary odors, (iii) statistical

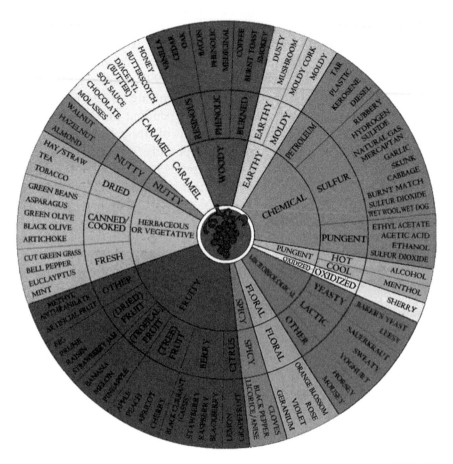

Fig. 4.2. The classification of olfactory families for wines. Reprinted with permission from Ann Noble, 1990. © Wine Aroma Wheel copyright 1992, 2002 A.C. Noble, www.winearomawheel.com.

descriptors supported by odor profiles and semantic descriptions, (iv) similarity data, (v) quantitative structure–activity relationships (QSARs) and olfactophore models, and, more recently, (vi) biochemistry and neuronal activity patterns. More detailed insights on these topics can be found in the literature (Dravnieks, 1966; Wells and Billot, 1988; Callegari et al., 1997; Chastrette, 1997, 2002; Wise et al., 2000; Mamlouk et al., 2003; Mamlouk and Martinetz, 2004; Teixeira et al., 2010, 2011).

First and foremost, it must be noted that any attempt to establish comparisons between these classifications must be cautious since they were built with different purposes and/or are supported by distinct

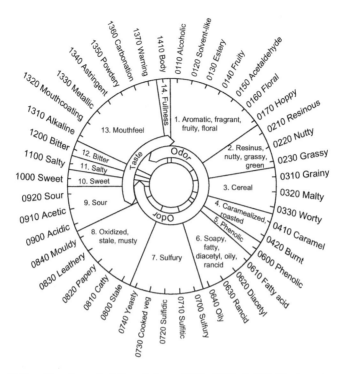

Fig. 4.3. The Meilgaard beer flavor wheel. From Artiga and Milla (2009). © 2009, Elsevier.

theoretical approaches. As a general trend, it is seen that the oldest classifications of odors often presented a small number of olfactive families as opposed to recent ones which tend to increase the number and diversity of quality descriptors. Moreover, odorants may be classified into one family only, while in other cases they may be assigned to subfamilies or *nuances*, as we have seen before.

1. *Empirical classifications*

As expected, the first attempts to classify odors into olfactory family classes relied on empirical classifications. These are mainly quantitative (although may be argued that they are also very subjective) classifications which are based on the olfactive ratings given by panelists (usually experts). The first one trying to establish a unified classification of odors was probably Aristotle, who considered six classes, two pleasant (*fragrantes* and *aromatici*), two as fetid (*tetri* and *nauseosi*), and two classes as pleasant for some and unpleasant for others (*ambrosiaci* and *hircini*). Later in 1756, Linnaeus proposed in his *Odores Medicamentorum* a modification to that classification by introducing another class of medicinal properties (*alliacei*).

However, the majority of empirical classifications came out from perfumers, following the development of the F&F industry in the end of the nineteenth century. There are dozens of different empirical classifications of fragrance raw materials, developed by experienced perfumers of the most popular fragrance companies and so supported on their olfactory expertise. Here, we can refer the odor descriptor wheel by St. Croix Sensory Company (McGinley et al., 2000) having 8 families and dozens of subfamilies and the Rosace of Firmenich from 1972 (Chastrette et al., 1991) with 13 families.

A detailed description of some examples of such classifications was given by Wells and Billot (1988). Back to the end of the nineteenth century, when Zwaardemaker, the first major olfactory psychologist, proposed a classification based on nine families—aromatic, ambrosiac, alliaceous, foul, fragrant, empyreumatic, ethereal, hircine, and nauseous—which were subdivided in 30 subfamilies (Zwaardemaker, 1927; Wise et al., 2000). As described by Chastrette (2002) other classification systems were proposed by Billot in 1948 with nine families, each with 2–10 subfamilies, and by Brud in 1986 with his odor ring with 12 families (animal, aldehydic-ozone, balsamic, citrus, conifer-evergreen, floral, fruity, green-herbal, lavender, mossy, spicy, and woody).

2. *Stereochemical theories and primary odors*
In the beginning of the twentieth century, Henning (see Wise, et al., 2000) proposed a semiempirical classification of primary odors using an odor prism in an attempt to elucidate the preconceived idea of "complex odors." Henning's prism comprised six corners labeled as putrid, fragrant, spicy, resinous, burned, and ethereal. However, experimental tests resulted in great variations in where the different odors should be placed on the prism, and so Henning's theory fell out of favor (Wise et al., 2000). Later, Crocker and Henderson (1927) proposed another semiempirical classification system based on four families: fragrant, acid, burnt, and caprylic (Ross and Harriman, 1949). They considered the endless combinations of these four scents would be responsible for

all odors we perceive. Although the simplicity of this classification may look attractive, perfumers considered this classification of odors too rudimentary.

Furthermore, it is important to mention the work of Amoore (Amoore et al., 1975; Amoore and Forrester, 1976; Amoore and Hautala, 1983) which represents a more theoretical view of olfaction. His stereochemical theory was based on assertion of the relationship between odor and molecular chemical structures. This theory led him to search for a limited number of discrete primary odor sensations based on specific anosmia studies that could express the sense of smell. It started to comprise seven primary odors (ethereal, camphoraceous, musky, floral, minty, pungent, and putrid) which were later extended to 25 though there may exist hundreds of them (Amoore et al., 1975; Amoore and Forrester, 1976; Chastrette, 1998; Weiner, 2006; Haddad et al., 2008). Among the controversies arising from the stereochemical theory, it should be noted the multiple examples of odorant molecules with very similar shapes but completely different odor character and/or intensity. Isomers and, in particular, enantiomers or molecules with certain different functional groups are among the examples for this case (e.g., exchanging hydroxyl and thiol groups) (Rossiter, 1996; Rowe, 2005). On the other hand, it is also common to observe that odorants with dissimilar structures have similar odor intensities and/or character (Sell, 2006). Another theory for primary olfactory perception was proposed by Luca Turin, who rejuvenated the older theories of Dyson and Wright, proposing that our nose and our olfactory receptors would be specific-responsive to the molecules' vibrations instead of to their shape (Turin, 1996, 2002, 2005). However, several objections have hindered further success of this theory especially when cases of optical isomerism and isotopic substitution are present in odorant molecules (Rossiter, 1996; Rowe, 2005). We will not go into further details on this topic but we recommend the reader to follow the literature.

3. *Statistical descriptors supported by odor profiles and semantic descriptions*
   The application of multidimensional statistical methods is often applied to large datasets attempting to correlate odor quality. Techniques used are typically principal component analysis (PCA) and multidimensional scaling (MSD). According to Chastrette (2002), there are three different types of data that can be used: classic

semantic descriptions, descriptions emphasizing similarities, and odor profiles for odor intensity estimation. Here, we will only highlight some of the works developed in this field: the comprehensive semantic odor-profile database from Arctander and Zarzo (Arctander, 1969; Sigma–Aldrich, 2003; Zarzo and Stanton, 2009); the set of odor profiles from Boelens and Haring for 309 chemical odorants (see Chastrette, 2002), the Atlas of odor character from Dravnieks in 1985 (Wise et al., 2000), the olfactory space developed in 1997 by Laffort et al. (Callegari et al., 1997), and, more recently, the perceptual space of Zarzo (2008) which tried to classify fragrances by structure-odor relationships (SORs) or similarity tests using extensive databases of descriptors (for further details, see Chastrette, 1997, 1998; Pintore et al., 2006; Haddad et al., 2008).

However, such types of classifications generally include mapping multidimensional maps of odor quality descriptors. Their lack of simplicity for consumer application turns them almost impossible to employ. On simple perceptual grounds, the lack of clarity for the interpretation of these multidimensional scaling analyses with n-dimensional spaces is a difficulty for their application by nonexperts.

4. *Similarity data*

There were also empirical classifications based on similarities between odors, though these are strongly dependent on the system used. The Field of Odors first presented by Jaubert et al. in 1987 (Jaubert et al., 1995) is an attempt to classify the olfactory space by similarities found by experts between pairs of odorants. Moskowitz and Gerbers developed a two-dimensional figure of odorants using similarity ratings among 15 odors and 17 descriptors. Although the list of authors studying this type of approach goes on, it is conclusively that such methodologies felt off favor because it is considered that classes of odors are sharply delineated as stated by Chastrette (Chastrette, 2002).

5. *QSAR and olfactophore models*

The so-called QSAR techniques are valuable tools for the F&F industry, mainly for the modeling of odor detection, intensity, and character, resulting in the so-called SORs. However, due to the complexity relying in odor perception, SOR models are often applied to distinct odor groups separately, such as ambergris, almond, musk, or sandalwood (Rossiter, 1996). The reasons why it remains difficult to apply SOR methods to all types of odorant molecules have to do with two points: first, odorant molecules that can be described with one or (at most) two semantic descriptors are easily classified; second, complex and flexible odorant molecules are more likely to adopt different energetically favorable conformations, each of which may trigger different odor responses (Rossiter, 1996). A detailed description of the application of SOR models is described in that work.

More recently, a novel application of SOR models has been devoted to what can be called as olfactophore models, in analogy with pharmacophore models widely used for the study of biological activity of drug molecule candidates. The idea relies on mapping a set of molecular properties of the odorant molecule that are considered to be critical for its odor. At the same time it attempts to model the odorant-receptor site binding properties and steric geometries (Kraft et al., 2000; Kraft and Eichenberger, 2003). Such models are based on the molecular similarity of known odorants. Kraft and Eichenberger (2003) developed the olfactophore model for the correlation of structure-odor properties for a small class of 20 marine odorants. However, the predictive capacity of the olfactophore models developed so far is still not enough to discriminate between odorant chemicals with very similar structures, which are known to produce very different olfactory qualities.

6. *Biochemistry and neuronal activity patterns*

Studies at the frontier of olfactory recognition have been looking for understanding of both the binding mechanisms between odorant molecules and olfactory receptors (ORs) as well as on tracking and decoding neuronal activity patterns. Since the discovery of the large family of ORs from the Nobel Laureates Linda Buck and Richard Axel (Buck and Axel, 1991), this study has been focused on the binding interactions between ORs and odorant molecules (Spehr and Munger, 2009; Yabuki, et al., 2010). It has been shown that such interactions are the starting point for biochemical processes on odorant recognition and that the large and complex sets of ORs are responsible for the chemosensory ability of the olfactory system.

Another approach for odor recognition and classification lays on the study of neuronal activity patterns. Such evaluations consider spatial-temporal patterns of neural activity. Moreover, they consider that odor quality is recognized by the combination of neuronal fixed patterns which the brain classifies into discrete representations (Oyamada et al., 1997; Hoshino et al., 1998; Niessing and Friedrich, 2010). For further details, we recommend the reader to see the review on this topic by Jay Gottfried (Gottfried, 2007).

As a general conclusion, in the past years we have enlarged the available database of odors and chemical structures and learned in a more detailed level the mechanism between odorants and ORs. However, we are still far away from understanding their exact structure, and the way signal transduction and interpretation are processed (especially for character recognition), which limits any deeper interpretation of the olfactory mechanism (Rowe, 2005). At this point, it is important to highlight that common terms are found in many of these perfume classification systems (as well as some differences) suggesting that a universal system for fragrance classification may be reasonable (Milotic, 2003).

In this work, four databases were selected for the assignment of olfactory families to single fragrances (which combine types 1, empirical classifications, and 3, statistical descriptors supported by odor profiles and semantic descriptions, both for pure compounds). The literature databases of Brechbill (2006), Surburg and Panten (2006), The Good Scents Company (2010), and an in-house developed compilation of olfactive families from several perfume companies were considered, including more than 2000 fragrant species. However, it should be mentioned that the Brechbill's database was the first reference of choice while the remainder were used when the former had no classification assigned. Whenever more than one family was attributed, the most representative family was considered as primary family and the following classes were considered as subfamilies or *nuances*. Additionally, when there were discrepancies between the primary families assigned in different databases, the criterion was to consider the families of the Brechbill's database and then include as subfamilies the classifications of the remaining databases. The relative distribution of each family or descriptor according to the Brechbill's dataset is represented in Fig. 4.4. It is readily apparent that the floral family is the one that holds the largest share of the distribution among the fragrance raw materials (21%), not overlooking that the families named "Rose" (11%) and "Jasmine" (3%) also represent scents of flowers and are not negligible. This fact mirrors what is observed in commercial perfumes,

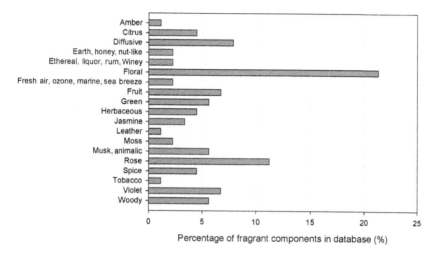

Percentage of fragrant components in database (%)

Fig. 4.4. Relative distribution of each olfactory family or descriptor in the database used for this work. Adapted with permission from Teixeira et al. (2010). © 2010, American Chemical Society.

where the number of floral fragrances is very significant, especially in women's perfumes.

4.2 CLASSIFICATION OF PERFUMES

So far, we have discussed the classification of single odors. We agree with other authors' opinions that all odors are represented in the brain as complex patterns whether the odorant is a chemically pure substance or a complex mixture. Nevertheless, the recognition of single odors is easier than the recognition of odorants within a mixture (especially for nontrained people). Consequently, the classification of single (pure) odorants remains simpler than for mixtures, which may elicit several *nuances* or background odors. Apart from that, it is clear that the odor space of perfumes and most fragranced products will be a mixture of odorant chemicals which will certainly elicit different odor intensities and/or qualities. Thus, the olfactory classification of such products, although complex, remains elementary.

In what concerns to perfumes, several attempts have been made in the past trying to associate these complex mixtures to olfactory families. Such type of classification may be called as fragrance genealogy. Perfumes generally have around 50—100 fragrant components in their formulation, with several functional groups within this high number of molecules (hydroxyl, carbonyl, ether, and many others). Due to this complexity, they are often qualitatively classified into olfactory families attending to the dominant fragrant notes. Although the F&F industry employs sophisticated analytical tools in perfumery (e.g., GC/MS, olfactometers, and electronic noses), so far there is not yet a standard classification of perfumes in view of their olfactory nature (perfume families). This procedure is performed by perfumers who empirically attribute the perceived odor mixture to a class or family (floral, citrus, chypre, oriental, etc.). In such evaluation, there may be subfamilies or *nuances* present at lower intensities as well. Thus, it is mainly experimental, person-sensitive, and, perhaps most important of all, it is only applied in the postformulation step. We highlight this fact because it is behind one of the reasons why it is necessary to formulate dozens, if not hundreds, of perfume mixtures, until the desired scent is obtained. Additionally, even among these experts in the art of perfumery there are biological differences in their olfactory response to fragrance stimuli, resulting in different classifications (Jinks and Laing, 2001; Gilbert, 2008). On the other hand, it is important to point out that the classification of perfumes could also be performed by consumers that have little to no experience in the art of perception of odors. In 1992, Jellinek tried to follow this hypothesis and presented a consumer-based classification for perfumes in its "Map of the world of fragrances" and "Odor effect diagram" (Milotic, 2003).

In what concerns to empirical classifications of perfumes, some of them are available in the literature while others remain proprietary, most of which were developed by F&F companies. One of the most precious works, resulted in an extensive compilation from the survey of all perfumed products developed since 1782, which lead to the *Classification des Parfums et Terminologie* by the French Society of Perfumers (1984) (Société-Française-des-Parfumeurs, 2006). It was divided into five main olfactory families, plus several secondary families or *nuances*. Later on, a second study outlined in 1989 included the contribution of two new families, citrus and woody, and the addition of secondary families (Société-Française-des-Parfumeurs, 2006). Another valuable and extensive classification is the fragrance wheel developed by the acknowledged perfumer Michael Edwards (first presented in 1983). It considers four standard family notes, each one having three subfamilies (Edwards, 2009), comprising 14 primary families as shown in Fig. 4.5. Edwards has already classified more than 5700 commercial perfumes (and more than 6500 fragrance products) using these 14 categories displayed

Fig. 4.5. The fragrance wheel from Michael Edwards: first developed in 1983, suffered updates over the years (latest version in 2012). Reprinted with permission from Edwards (2011). © 1992–2012 Michael Edwards.

around a central hub, except for aromatic/fougère, which in some cases is placed at the center of the wheel as it will be discussed later (Zarzo and Stanton, 2009). The ancestral fragrance wheel has also evolved over the years, mainly in what concerns the positioning of the olfactive families around the central hub. Zarzo and Stanton have greatly contributed to that by finding consistencies between the odor effect diagram of Jellinek and the wheel of Edwards. From their studies, they have proposed the replacement of fruity family between floral and green, while water would change from its previous location to between green and citrus (Fig. 4.5). They also suggested that the original aromatic/fougère central hub could be placed near dry woods and citrus which Edwards accepted (in fact, he already did that for perfumers while for retailers and customers this classic men family was presented in the center of the wheel) (Donna, 2009).

Another very good classification map for fragrances or perfumes is the "Drom fragrance circle" (DROM, 2011) shown in Fig. 4.6. It considers 16 olfactive families classified by perfumers (outside band) and consumers (inner band) related terms. In the middle band, some examples of essential oils are given for each family. Moreover, in the center

Fig. 4.6. Drom fragrance circle: different layers are used to relate the classifications of essential oils using perfumers related terms and those from consumers. Reprinted with permission. Copyright 2012 by DROM Fragrances.

of Fig. 4.6 is shown an example for a classification of an essential oil where the intensity of the olfactory families is represented through five (in this case) colored intensity slices, thus generating a stellar map.

Besides, some fragrance companies like Avon (2009) classifies perfumes into six different families, each one comprising two subfamilies or *nuances*. The experimental classification of Osmoz by Firmenich (2012) has a six-group family criteria for both women and men with various *nuances* in each one. ScentDirect (ScentDirect, 2009), and H&R Genealogy by former company Haarman & Reimer (H&R, 2002) divide their classification in six and four families with different subfamilies, respectively. Octagon, by former company Dragoco, now merged with H&R into Symrise, has nine perfume families. It is curious that there is a more detailed description of the floral family in this classification differentiating between simple and complex floral accords. Finally, a compilation study on thousands of commercial perfumes made by

Luca Turin and Tania Sanchez (LT&TS), a recognized scientist and a perfumer, classifies fragrances into a large number of olfactive families, always giving a touch of their personal assessments about them (Turin and Sanchez, 2008). However, it should be mentioned that in some cases the definition of the correspondent family was somewhat ambiguous. Finally, it is to be said that other fragrance classifications have been proposed but in most cases, their details still remain confidential.

In this work a new approach for perfume classification has been developed. The aim of the PR methodology is to predict the classification of perfumes into olfactive families as performed by perfumers using physicochemical models and qualitative descriptors. The methodology combines the use of radar graphs (to present qualitative information) with some Product Engineering tools and concepts previously developed (Teixeira et al., 2010). The Osmoz, ScentDirect, H&R Genealogy, Octagon, The Fragrance Foundation, and LT&TS perfume classifications were used for comparison with the proposed methodology for the classification of perfumes. All the olfactory families considered in these classifications along with that of the authors (Teixeira et al., 2010) are presented in Table 4.1.

4.3 THE PERFUMERY RADAR (PR) METHODOLOGY

The PR methodology is a software tool with predictive capabilities for the classification of perfumes into olfactive families. In simple words, it allows a qualitative description of the perceived odor elicited from any fragrance mixture. At the outset, the added value of the PR methodology is centered in two topics: firstly, it introduces some scientific basis in fragrance classification, reducing the arbitrariness to the empirical classification of pure odorants; secondly, it is a predictive tool and so it can be helpful in the preformulation stages of fragrance development, helping to reduce time and cost of product.

It is a stepwise methodology that combines different models (for VLE, odor intensity, or odor character) in a software tool as structured below:

1. Classification of pure fragrance ingredients into olfactory families;
2. Prediction of the composition in the gas phase and odor intensity for each component;
3. Calculation of the odor intensity for each olfactory family and graphical representation of the PR.

Table 4.1. Perfume Families Used in Different Perfume Classifications

	FSP[a]	FSP[b]	AVON[c]	TFW[d]	Osmoz[e]	SD[f]	H&R[g]	Octagon[h]	This Work
Citrus		✓	✓	✓	✓				✓
Floral	✓	✓	✓	✓	✓	✓	✓		✓
Green				✓					✓
Fruity				✓					✓
Herbaceous									✓
Musk									✓
Oriental			✓	✓	✓	✓	✓	✓	✓
Woody		✓	✓	✓	✓			✓	✓
Chypre	✓	✓	✓		✓	✓	✓	✓	
Fougère	✓	✓	✓			✓	✓	✓	
Leather	✓	✓							
Amber	✓	✓							
Fresh				✓					
Dry Woods				✓					
Floral-Oriental				✓			✓		
Mossy Woods				✓					
Soft-Floral				✓					
Soft-Oriental				✓					
Water				✓					
Woody-Oriental				✓					
Aromatic					✓				
Aromatic-Fougère				✓					
Cedar							✓		
Aromas								✓	
Eau de Cologne								✓	
Floral Simple				✓				✓	
Floral Transparent				✓				✓	
Floral Bouquet				✓				✓	

[a]French Society of Perfumers (1984)
[b]French Society of Perfumers (1989)
[c]AVON.
[d]The fragrance wheel by Michael Edwards in The Fragrance Foundation (1983).
[e]Osmoz by Firmenich.
[f]ScentDirect (SD).
[g]Haarmann & Reimer (H&R).
[h]Octagon by Dragoco.
Source: Adapted with permission from Teixeira et al. (2010). © 2010, American Chemical Society.

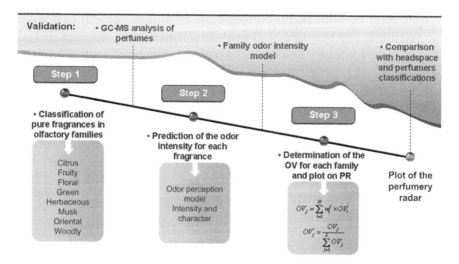

Fig. 4.7. Schematic representation of the main steps for the PR methodology: on the top are presented the validation steps and on the bottom the steps for the prediction of the PR plots. Adapted with permission from Teixeira (2011). © 2011, M.A. Teixeira.

The schematic representation of the steps followed in the PR methodology along with the procedures for its validation is shown in Fig. 4.7. The validation of the PR is crucial for its application as a predictive tool. In this way, two different validations were performed for a series of essential oils and commercial perfumes: on one side, a comparison was done with the classifications from perfumers, which are based on their olfactory evaluations; while on the other hand, a validation of its predictive capabilities was performed with an experimental PR obtained by measuring the headspace compositions of fragrance mixtures.

Step 1—Classification of pure fragrance ingredients into olfactory families

In what concerns the PR methodology, its first step comprises the classification of pure fragrance ingredients into olfactory families and, then, the selection of the major olfactory families we want to work with. It is known that different fragrance houses use different classifications. Thus, despite the modular characteristic of the PR, it makes sense to choose the most relevant olfactory families in order to obtain a well-distributed radar. These olfactory families may be those that are most frequent in the classification of pure components (e.g., floral as seen in Fig. 4.4), or may be olfactory families that result from the combination of different *nuances* but have a great relevance in the world of perfumery (e.g., chypre). In what concerns the classification of pure fragrance ingredients, it was used in this work the extensive compilation of Brechbill (2006) and it was decided to include three olfactory descriptors for each compound (e.g., limonene is classified as fresh-citrus-orange). Taking into account these data and the literature olfactory classification data from F&F companies, eight olfactory families (citrus, floral, green, fruity, herbaceous, musk, oriental, and woody) were selected for the PR. The main criterion was based on the most commonly used terms for the classification of pure fragrances. The selected eight olfactive families are briefly described below (Poucher, 1955; Calkin and Jellinek, 1994; Butler, 2000; Brechbill, 2006; Edwards, 2009; Firmenich, 2012; Leffingwell and Leffingwell, 2012):

1. *Citrus*: freshness and lightness from citrus fruits like lemon or orange. The first *Eau de Cologne* ever made belonged to this family.
2. *Fruity*: from natural fruits like apple, banana, or raspberry.
3. *Floral*: made up of flowers (e.g., geranium, jasmine, or rose) is one of the most widely used families for feminine fragrances.
4. *Green*: typical botanical notes with scents of fresh leaves or stalks and mown grass or with reminiscent freshness (e.g., vertocitral, hexenyl benzoate).
5. *Herbaceous*: more complex scents than green, often found in low-growing plants. Typical examples are sage and mint.
6. *Musk*: characteristic from the musk deer and musk oxen. The odorants of this family when used in perfumes often act as fixatives (components that fix other fragrances in the solution).
7. *Oriental*: associated to amber species, often including warm scents. The classifications found in the literature for scents like spicy, earth, balsamic, tobacco, leather, waxy, and mossy were included in this family.
8. *Woody*: generally as woods like cedarwood, sandalwood, or patchouli. A classification of fragrances as camphoraceous was included in this family.

Moreover, positioning the families around the radar axis must take into account the similarities and differences between olfactory notes within perfume formulation. For example, if we recall the fragrance wheel from Michael Edwards (Fig. 4.5) or the Drom fragrance circle (Fig. 4.6), both have placed their olfactive families around the circle following their own perspective of similarity and blending. Nevertheless, it is also obvious from Figs 4.5 and 4.6 that this distribution of olfactive families is completely different in the two diagrams, thus showing, once more, the large discrepancy in fragrance classification. In our case, it took into account the traditional subfamilies or *nuances* existing in each family, so closer axes usually represent families that blend well in perfumery products.

Step 2—Prediction of the composition in the gas phase and odor intensity for each component

Then, in the second step, the composition (x_i, mole fraction of each fragrant ingredient) of several commercial perfumes was determined by gas chromatography coupled with flame ionization detector (FID) and mass spectrometry (GC/FID/MS) to identify and quantify single fragrant components. Because throughout this analysis water molecules cannot be detected, a specific amount of water (solvent) was considered depending on each type of perfume (*Eau de Parfum, Eau de Toilette, Eau Frâiche*). The mixture composition was then normalized for each perfume. Nevertheless, a parenthesis is made here, to highlight that for the perfumer this analytical evaluation is not needed because he or she knows the exact formula of the fragrance. Having determined the composition in the liquid perfume, the OV for each fragrance ingredient is calculated using Eq. (2.4) as shown in Chapter 2. The same methodology as that used in the PTD® and PQ2D® is followed here: the UNIFAC method is used to estimate the activity coefficients (γ_i) in the liquid phase, and the OV is applied to calculate the odor intensity of single components. Of course, several physicochemical and psychophysical properties (molecular weight, vapor pressure, UNIFAC interaction parameters, and odor detection threshold) have to be compiled for all the fragrant compounds existing in the perfume. It should be mentioned that the OV concept was preferred in detriment to the well-known Stevens' Power Law (Stevens, 1957) due to the larger number of fragrance ingredients present in a perfume and, consequently, the difficulty in obtaining Power Law data (the number of published exponents is less than 10% of the number of existent fragrant compounds). Nevertheless, if this parameter is known or experimentally measured for all compounds, the Power Law can perfectly be applied in the PR methodology.

Step 3—Calculation of the odor intensity for each olfactory family

In the third step, the OV for each fragrance ingredient and, consequently, for each olfactory family (summation of the OVs of all the fragrance ingredients) is calculated. For that purpose, once literature classifications of pure ingredients differ or often attribute more than one family to a compound, olfactory families were considered as defined by Brechbill (2006) and whenever other families were also referred, these were used as secondary (and tertiary) families according to the order of appearance or relevance given. Taking that into consideration, a weighing criterion was used in order to account for the presence of *nuances*. This weighed olfactory family intensity model considers the maximum of three subfamilies per fragrance ingredient as presented in Table 4.2.

This model considered that the primary family is more strongly perceived than the secondary family, and this one is more intense than the tertiary family. The reason behind the choice of three families or subfamilies lays on experimental studies that show, for pure substances, an average value of three descriptors per odor is considered sufficient. Indeed, Chastrette (1998) obtained an average of 2.8 different descriptors for thousands of odorants. In this way, the quantification of the OV for each olfactory family is calculated by:

$$OV_j = \sum_{i=1}^{N} w_i^j \times OV_i \qquad (4.1)$$

where w_i^j is the weight factor of component i for family j according to Table 4.2. It should be highlighted that the selection of the weight factors was arbitrary, considering the more intense perception of the primary over the secondary family, and that over the tertiary. Moreover, these weights may vary from odorant to odorant but the fragrance classifications do not present any quantitative information concerning that point, and the same weight factors were used for all fragrances. It should also be noted that the sum of the OVs for the quantification of the family intensity is an approximation considered

Table 4.2. Distribution of Weights (w_i) for Each Olfactory Family

Number of Families	Family		
	Primary	Secondary	Tertiary
1	100%	–	–
2	70%	30%	–
3	60%	30%	10%

Source: *Adapted with permission from Teixeira et al. (2010). © 2010, American Chemical Society.*

in our model. By doing this, effects like hyper- or hypoadditivity of odors (the perceived intensity of an odorant mixture to be higher or lower than the sum of the perceived odor intensities of the single components) are not considered in this methodology (Laffort and Dravnieks, 1982; Olsson and Berglund, 1993; Cain et al., 1995). Other particular olfactory effects (e.g., change of odor quality with concentration, enhancement of the intensity of a strong odor by a weaker odor) are also out of the scope of the model since its purpose is to obtain a qualitative analysis in terms of olfactory families (Laffort and Dravnieks, 1982; Fisher and Scott, 1997).

Finally, since the PR diagrams are independent of the total odor intensity, these olfactory family OVs (OV_j) are normalized according to:

$$OV'_j = \frac{OV_j}{\sum\limits_{j=1}^{L} OV_j} \tag{4.2}$$

where OV'_j is the normalized OV for family j, and L is the number of olfactory families defined (in this work $L = 8$). By doing this transformation, it is possible to compare all the PRs in a scale independent of the odor intensity. Finally, the relative odor intensity for each olfactory family is represented using radar plots. This representation will show not only the dominant olfactory family but also its main *nuances* (if any) for the selected perfume.

As a proof of concept, the validation of the obtained radars can be performed in two ways: using the classifications from the F&F industry or by developing experimental PRs. In this way, the comparison with perfumers' evaluations is a straightforward task which can be achieved by direct analysis of their classifications. On the other hand, for the determination of the experimental PRs, the headspace of a liquid sample of perfume is evaluated by GC/FID/MS at equilibrium conditions. Once the vapor composition is known, Steps 2 and 3 of the PR methodology are applied. However, it should be noted that perfumes are complex mixtures with hundreds of chemical odorants. Thus, it is no surprise that we find tens or hundreds of odorants in their headspace. For simplification of this validation process, the FID data was reduced to a lower-limit value of peak area ($<10,000$ counts) in the integration to exclude components or residues present in the analysis of the perfumes. It is known that at very low-peak areas the noise-to-signal ratio in FID detectors is high, thus introducing errors in the composition. Moreover, the peak identification performed by mass spectrometry turns to be more difficult whenever the chromatographic peaks are too small. Additionally, accounting for peaks much below this limit (e.g., 1000 counts) would significantly increase the number of components and so make its identification too laborious and almost impractical. Nevertheless, care should be taken with this rejection limit, once powerful odorants may be present in trace amounts (have small chromatographic peak areas) (Ohloff, 1978). After knowing the gas phase composition of the perfume, steps 3 and 4 of the PR methodology are performed and the experimental radar is obtained.

The experimental validation of the PR methodology was first done using simple and well-known essential oils, and later with commercial feminine and unisex perfumes already classified into olfactory families by experienced perfumers. The former are simple mixtures with 30 or less fragrant compounds in their composition, while the later (the perfumes) are more complex having 50–100 or more. The validation of the predictive capability of the PR methodology was also done by analyzing the gas phase above the perfume (headspace) in equilibrium conditions using GC/FID/MS and then calculating the PR from real (not predicted) gas compositions.

4.4 APPLICATIONS OF THE PR METHODOLOGY

First, we will show the application of the PR methodology to four essential oils commonly used in the F&F industry: orange, lemon, jasmine, and thyme. By doing this, it is possible to have a preliminary idea of the potentialities of the PR methodology. Once this step is verified, then it will be the time to apply the PR methodology to more complex mixtures, such as commercial perfumes.

4.4.1 Essential Oils

The classification of the essential oils into olfactive families is presented in Table 4.3 and the corresponding PRs are shown in Fig. 4.8. According to the literature, both the essential oils of orange and lemon belong to citrus olfactory family (Brechbill, 2006). Accordingly, we can see that the two predicted radars in Fig. 4.8 are very similar in terms of the main olfactory families and their relative intensities, as expected. Nevertheless, it must be pointed out that although these

Table 4.3. Olfactive Families Given by the Classification of the Brechbill Database for the Essential Oils of Orange, Lemon, Jasmine, and Thyme (Brechbill, 2006)	
Essential Oils	**Family**
Orange	Citrus
Lemon	Citrus
Jasmine	Floral
Thyme	Herbaceous

Source: *Adapted with permission from Teixeira et al. (2010). © 2010, American Chemical Society.*

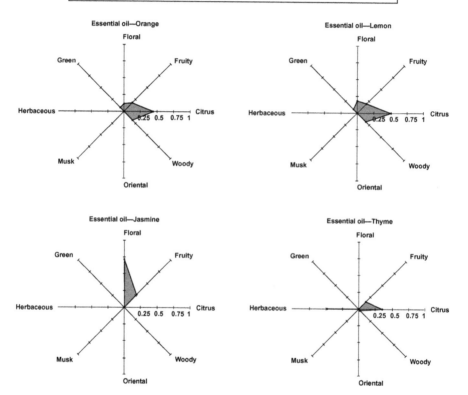

Fig. 4.8. PRs for the essential oils of orange, lemon, jasmine, and thyme. Adapted with permission from Teixeira et al. (2010). © 2010, American Chemical Society.

two essential oils may be derived from citric fruits, their olfactory space will present minor differences. In fact, their composition (both liquid and vapor) are similar but there are some odorants in one essential oil and not in the other, making the perceived scents different. From the PRs, it can be seen that the dominant olfactory family for them is citrus, with 44% and 49% for orange and lemon oil, respectively. Such result makes reasonable sense, since as known from everyday experience, lemon has a stronger citrus character than orange. If we take a look at the subfamilies, it is seen that woody-fruity-floral *nuances* are predicted for the orange essential oil while for lemon there is a fruity-floral-woody character. Moreover, a green character is perceived in the orange but not in the lemon essential oil. These differences in the type and relative intensities of the olfactory families mean that there will be certain differences in their perception.

The other two essential oils evaluated with the PR methodology are classified as classic floral family for jasmine oil while thyme exhibits a characteristic herbal scent. For the first, its PR gives a dominant floral scent (71%) with a fruity *nuance* (27%) and the corresponding radar for thyme oil shows that the herbaceous family dominates (48%), together with *nuances* of citrus (34%) and fruity (14%).

Thus, from the PRs of four widely used essential oils, it is possible to observe that the predicted primary olfactive family is always in agreement with the classifications from perfumers. Additionally, some subfamilies or *nuances* are also predicted for their perceived odor, showing minor differences between very similar essential oils.

4.4.2 Commercial Perfumes

A curious point in the fragrance industry that has not yet been discussed here concerns to one other type of fragrance classification: the classification in gender. In analogy with the sense of smell, studies on the visual sense using semantic differential scales have shown that pink, yellow, and purple colors are considered more feminine, while blue, brown, and gray are most commonly classified as masculine (Zellner et al., 2008). We all know that some fragrances are typically more feminine whilst others show a more typical masculine scent. However, this distinction has mostly to do with fashion trends and personal and cultural preferences. Women's perfumes tend to be lighter and floral while masculine ones often have a stronger character of a citrus-woody type. Furthermore, apart from these two genera, there still exists (especially more recently) a line of unisex perfumes which aims at pleasing both the feminine and the masculine side. These differences depend on the composition of the perfume and are clearly evident in the main olfactory families in which they fall under. A comparison of perfume classifications in terms of olfactive families distributed by gender and type of notes is presented in Table 4.4 considering a compilation of 5233 fragrances (3149 women, 1241 men, and 843 unisex) contained in the 2012 edition of Michael Edwards' guide. It should also be highlighted that the number of feminine fragrances considered is much larger than that for men (as in line with market), mainly because the latter is a seasonal market (with its busiest period in the Christmas season), while the women's sector covers equally the entire year.

According to Zarzo and Stanton (2009), the analysis on the trend of each family can be performed in a simple way by considering that families whose percentage is higher for women tend to be feminine or gynogenic (perceived with female characteristics), whereas if it is greater for men then it is regarded as masculine or androgenic. Following the same analysis, it is possible to say that families with similar number of perfumes (equally distributed) may be associated with unisex fragrances as it is the case for aromatic, woody-oriental or citrus, although there are exceptions. Unisex fragrances are not perceived strictly as masculine or feminine. In this way, it seems that olfactory families like citrus, woody-oriental, or woody are in the boundary between femininity and masculinity. For example, floral-oriental or fruity are feminine descriptors and contain (almost) only women's and unisex fragrances.

Another important family to evaluate is the mossy woods which corresponds to the chypre family (Zarzo and Stanton, 2009). Here, there is a slight discrepancy that deserves our attention: Table 4.4 shows that in the mossy woods family there are more feminine than masculine fragrances, so it would be gynogenic, but according to the H&R guide, there are 23.8% of feminine fragrances and 35.7% of masculine fragrances classified as chypre, showing that it is more androgenic. This reflects once more the difficulty in the classification of "more complex" scents (like the chypre family) by the human nose, even when experts' opinions are used. However, it is clear that all these analyses and discussions are not sufficient to define what the ideal perfume is for a woman or for a man.

A recent study performed by the Fragrance Foundation showed that age is also an important factor to take into account in this matter. Through a survey, they concluded that women under 18 have a higher propensity for citrus fragrances, whilst in the 18−25 range they prefer

Table 4.4. Percentage of Fragrances by Category/Descriptors Commonly Used in Perfumery and Gender Using as Database the 2012 Michael Edwards' Guide

Olfactory Family	Number of Fragrances			% of Fragrances by Gender			% of Olfactory Family	% of Fragrances by Gender in Olfactory Family		
	Women	Men	Unisex	Women	Men	Unisex		Women	Men	Unisex
Fruity	75	0	14	2.4	0.0	1.7	1.7	84.3	0.0	15.7
Green	30	7	24	1.0	0.6	2.8	1.2	49.2	11.5	39.3
Water (marine)	34	76	31	1.1	6.1	3.7	2.7	24.1	53.9	22.0
Floral	1331	9	112	42.3	0.7	13.3	27.7	91.7	0.6	7.7
Soft-Floral	282	12	40	9.0	1.0	4.7	6.4	84.4	3.6	12.0
Floral-Oriental	436	2	16	13.8	0.2	1.9	8.7	96.0	0.4	3.5
Soft-Oriental	98	8	39	3.1	0.6	4.6	2.8	67.6	5.5	26.9
Oriental	97	28	58	3.1	2.3	6.9	3.5	53.0	15.3	31.7
Woody-Oriental	300	256	96	9.5	20.6	11.4	12.5	46.0	39.3	14.7
Woods	153	284	128	4.9	22.9	15.2	10.8	27.1	50.3	22.7
Mossy Woods	161	52	38	5.1	4.2	4.5	4.8	64.1	20.7	15.1
Dry Woods	39	90	69	1.2	7.3	8.2	3.8	19.7	45.5	34.8
Citrus	106	103	160	3.4	8.3	19.0	7.1	28.7	27.9	43.4
Aromatic	7	314	18	0.2	25.3	2.1	6.5	2.1	92.6	5.3
Total	3149	1241	843							

fruity-floral notes, and above 35 years of age women mostly choose musky notes (TFF, 2011).

Thus, one concludes that there is still a long way to go in the classification of odors and their tendency toward women or men is difficult to predict *a priori*.

4.4.2.1 Feminine Fragrances
In what concerns the feminine class, 14 commercial perfumes were selected and evaluated with the PR methodology. Table 4.5 shows their commercial name, brand, and olfactive family classifications according to six different companies/authors together with the PR classification. As previously noted, it is important to remember that the former classifications are based on empirical evaluations made by perfumers of the corresponding company while the PR classification is based on our prediction software.

Among the perfumes listed in Table 4.5, there are clear differences as well as similarities in the olfactive families. More relevant than the fact are the great divergences amongst the classifications from different companies for the same perfume (e.g., P12–P14), although there are also good agreement for others (e.g., P1–P4). The most significant discrepancies occur for secondary and tertiary olfactive families.

The PRs simulated at initial times ($t = 0$ s) for these feminine commercial perfumes are presented in Figs 4.9–4.11. These were grouped in terms of their primary olfactive families: Floral (P1–P4), Oriental (P5–P7), and Chypre (P8–P11), respectively. Finally, those with heterogeneous classifications are presented in Fig. 4.12.

The typical floral perfumes considered in this study (see Table 4.5, P1–P4), produced PRs (Fig. 4.9) with a predominant floral scent which represents always more than 50% of the total odor (in terms of relative odor intensity). Additionally, the PR for P1 revealed a significant contribution of the green subfamily, which surprisingly is only described by the classification of LT&TS (Table 4.5). Nevertheless, the green olfactive family is also closely linked with the floral one, as can be seen by their proximity in the PR (or, e.g., in the fragrance wheel of Edwards). The classification of P1 from Dragoco and Institut Supérieur International du Parfum de la Cosmétique et de l'Aromatique Alimentaire (ISIPCA) is floral bouquet, which resembles different types of botanical flowers

Table 4.5. Commercial Name, Brand, and Family Classification of the Selected Perfumes

No.	Perfume	Brand	Osmoz (Firmenich)	ScentDirect	H&R	The Fragrance Foundation	Dragoco	LT&TS	Perfume Intelligence	ISIPCA	This Work
P1	L'air du Temps	Nina Ricci	Floral-Spicy	Floral	Floral	Floral	Floral Bouquet	Floral-Green	Floral	Floral Bouquet	Floral-Green
P2	Paris	YSL	Floral-Rose Violet	Floral	Floral	Floral	Floral-Rose	Rose	Floral	Floral Bouquet	Floral
P3	Chanel 19	Chanel	Floral-Green	Floral-Green	Floral-Green	Soft-Floral	Floral-Green	Green-Floral	Green-Floral	Floral-Green	Floral
P4	Eau de Givenchy	Givenchy	Floral-Fruity	Floral-Fruity	Floral-Fruity	Floral	Floral Transparent	Green-Floral	Green-Floral-Fruity	Floral-Green	Floral
P5	Addict	Dior	Oriental-Vanilla	Oriental-Floral	–	–	–	–	Floral-Oriental	Oriental-Floral-Woody	Oriental-Woody
P6	Addict EdF	Dior	Oriental-Floral	–	–	–	–	–	–	–	Oriental-Fruity-Green
P7	Gloria	Cacharel	Oriental-Woody	Oriental-Fresh	–	–	–	Amber-Rose	Floral-Oriental	Oriental-Floral-Woody	Fruity-Oriental
P8	Eau de Rochas	Rochas	Citrus-Aromatic	Chypre-Fresh	Chypre-Fresh	–	–	Citrus-Woody	Citrus-Chypre	Citrus-Floral-Chypre	Chypre
P9	Ô de Lâncome	Lâncome	Citrus-Aromatic	Chypre-Fresh	Chypre-Fresh	–	–	Fresh-Citrus	Fresh-Citrus	Citrus-Floral-Chypre	Chypre
P10	Miss Dior	Dior	Chypre-Floral	Chypre-Floral	Chypre-Fruity	–	Chypre-Green	Dry Chypre	Fresh-Green-Woody	Chypre-Fruity	Chypre-Green
P11	Ma Griffe	Carven	Chypre-Floral	Chypre-Floral	Chypre-Animalic	Mossy Woods	Chypre-Fruity	Green-Chypre	Green-Chypre-Floral	Chypre-Floral	Floral-Green
P12	Jungle Tigre	Kenzo	Oriental-Spicy	–	Chypre-Fruity	–	Floral-Fruity	–	Fresh-Fruity-Floral	Oriental-Floral-Woody	Floral-Green
P13	CK One	Calvin Klein	Citrus-Aromatic	Chypre-Fresh	Chypre-Fresh	–	Floral Transparent	Citrus	Fresh-Citrus	Citrus-Floral-Woody	Woody-Fruity
P14	Le Feu d'Issey Light	Issey Miyake	Floral-Woody-Musk	–	Oriental-Spicy	Soft-Floral	Woody-Spicy	Milky Rose	Floral-Woody	Floral-Woody	Green-Floral-Musk

Source: *Adapted with permission from Teixeira et al. (2010). © 2010, American Chemical Society.*

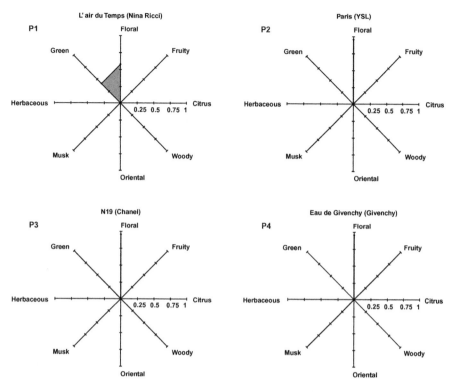

Fig. 4.9. PRs for perfumes P1—P4 (Floral primary olfactive family). Adapted with permission from Teixeira et al. (2010). © 2010, American Chemical Society.

and so reminds of a floral-green character (Sturm and Peters, 2005). For perfumes P2—P4, the PR resulted in the predominance of the floral olfactory family (>95%), and none of the *nuances* are noteworthy. This prediction of the main family is in agreement with the majority of the empirical classifications from F&F companies.

In what concerns to oriental perfumes (P5—P7), sometimes referred to as ambers, these are a blend of animal scents and vanilla, often mixed with exotic floral-spicy scents and woods. These perfumes have a typical oriental character, although with some gentle *nuances*. The obtained PRs for three of these perfumes evidence the same dominant oriental character, showing also some secondary and tertiary olfactory notes. In fact, for perfume P5, the oriental family (42%) dominates the overall scent, followed by secondary woody (17%), green (16%), and citrus (8%) *nuances*. Its *Eau Fraîche* version (P6) shows also an oriental character (65%), plus a fruity *nuance* (27%). The last perfume belonging

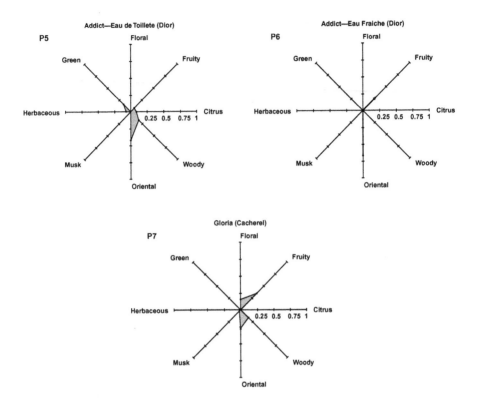

Fig. 4.10. PRs for perfumes P5–P7 (Oriental primary olfactive family). Adapted with permission from Teixeira et al. (2010). © 2010, American Chemical Society.

to this class (P7) is represented by a fruity-oriental character together with floral and woody *nuances*. This prediction partly agrees with the majority of the classifications from F&F companies which classify it as oriental and, in some cases, with different *nuances* like woody, floral, or fresh (e.g., fresh has been defined elsewhere as citrus-green by Surburg and Panten (2006) and citrus-green-fruity-water by Edwards (2009)). According to our PR methodology, it shows a fruity-oriental character with floral and woody subfamilies.

Another very important olfactory family within the perfumery business is the classical chypre. The word *chypre* owes its origin to the Mediterranean island of Cyprus, which for many centuries was the meeting point between East and West for trading of aromatic raw materials. Nowadays, however, it is related to a group of perfumes whose origins can be traced back to the great perfume Chypre, created by Francois Coty in 1917. The perfume is based on oakmoss and

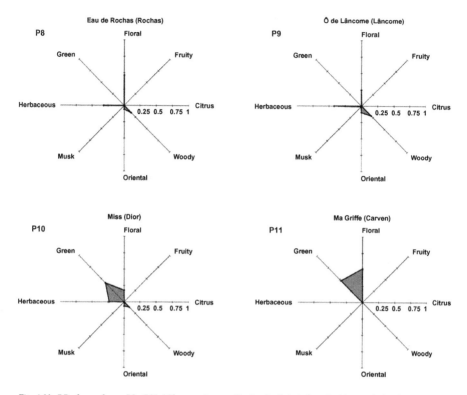

Fig. 4.11. PRs for perfumes P8–P11 (Chypre primary olfactive family). Adapted with permission from Teixeira et al. (2010). © 2010, American Chemical Society.

bergamot, together with jasmine, labdanum, and animal notes including civet and musk (Calkin and Jellinek, 1994). As we will see there is a great variety of chypre accords and/or perfumes. Such diversity led Calkin and Jellinek to assert that it *has led some perfumers to classify perfumes under three headings: florals, orientals, and all the rest, which may be generally described as chypres.* Despite this fact, we consider this to be relevant to apply the PR to such a class of perfume. In this way, Fig. 4.11 presents the PR for four perfumes (P8–P11) which are classified as chypre by the majority of the perfumers although some differences arise (Table 4.5). It should be noted that we do not have an axis in our PR methodology for this specific family, because of its complexity in the combination of odorants. In the literature, chypre is commonly described as a mixture of bergamot (fruity), rose (floral), patchouli (herbaceous-woody), and oakmoss (earthy, mossy, woody) (Calkin and Jellinek, 1994; Firmenich, 2012). Surprisingly or not, chypre family does not appear in the fragrance wheel of Edwards,

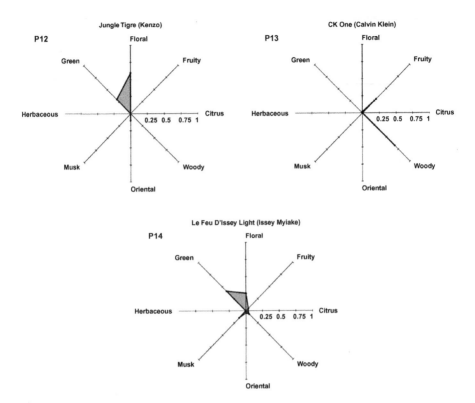

Fig. 4.12. PRs for perfumes P12–P14 (Heterogeneous classifications). Adapted with permission from Teixeira et al. (2010). © 2010, American Chemical Society.

and its placement would be difficult but it would be located under the oriental and woody families. For instance, Mitsouko (Guerlain), one of the classical chypre perfumes, is placed under mossy woods, but Rouge (Hermès), also a chypre perfume with a slight floral *nuance*, is placed under floral-oriental. Thus, it is expected a certain variation in the PRs for these perfumes (P8–P11).

From the radars in Fig. 4.11, it is possible to identify a pattern in three of these perfumes: PRs for P8, P9, and P10 present a similar shape having floral-herbaceous families as dominant notes, combined with woody-oriental *nuances*. In the case of P10, there is also a great contribution from the green olfactory family which is only described in the classification from Perfume Intelligence, thus evidencing an odor character slightly different. In the case of perfume P11, the PR reveals a clearly floral-green character which partly agrees with perfumers' classifications: floral is described as secondary family in four classifications while green appears as primary family in two. This green character can be compared with that from P10, though the herbaceous family is not present for P11. In fact, perfume P11 follows the description of LT&TS (Table 4.5) who classified this perfume as *a classic green-chypre, less herbaceous...more floral than most* (Turin and Sanchez, 2008). We have seen that this particular green odor is mainly due to the presence of styrallyl acetate (green-floral character) in the composition of the perfume. As a consequence, the ratio between its vapor pressure and its odor detection threshold is extremely high, resulting in a dominant OV for this component. After all, there is a similar shape in the PR of these perfumes belonging to the chypre family, although their formulation is quite different (both in species and composition). Thus, this radar shape can be assigned as a typical radar pattern for the chypre olfactory family, even though there may be different *nuances* within it.

Finally, the PRs for perfumes P12–P14, which present very different classifications from perfumers (heterogeneous classifications), are shown in Fig. 4.12. The discrepancies in the available classifications of these perfumes are so notorious in Table 4.5 that it gets hard to say what the correct classification is (if not impossible). Some of the primary olfactory families in some classification match some of the *nuances* in other. Seemingly, the use of a tailored in-house classification looks to be the fortunate choice for perfume classification. As previously mentioned, interpersonal differences at the level of the sense of olfaction might point toward such method but would not be possible to use a standard methodology such as the PR.

At first sight, the obtained PRs for P12–P14 present some of the olfactory families described by perfumers, although their dominance in the overall odor character may be completely different. In the case of P12 it shows a strong floral character (59%), together with green and oriental subfamilies, which match the classification given by Dragoco for the primary olfactive family. The PR for P13 classifies it as woody-fruity. These are families belonging to the chypre-fresh class of perfumes, as mentioned before. Consequently, this PR shows only some of the families and so, there may be some deviation. Finally, P14 presents a green-floral character due to the presence of mefloral (green-floral), with a musk *nuance* mainly due to the high composition of galaxolide, a sweet musk fragrance (Brechbill, 2006; Surburg and Panten, 2006). Both floral and musk families are present in the classifications of the perfumers (Osmoz, Fragrance Foundation, and LT&TS).

In brief, the application of the PR methodology to feminine perfumes has shown a good match for most of the primary olfactive families assigned in the empirical classifications (performed by perfumers). In some cases even the secondary olfactive families are well predicted, although there are also some outliers (Table 4.5).

4.4.2.2 Unisex Fragrances

The PR methodology was also applied to 11 unisex commercial fragrances from different brands. The selected perfumes and their classifications into olfactive families according to several fragrance houses and experts are presented in Table 4.6. These classifications were compiled from the literature and correspond to those from Osmoz (Firmenich), ScentDirect, H&R, The Fragrance Foundation, LT&TS, the Perfume Intelligence, and the ISIPCA.

These commercial unisex fragrances correspond to real perfumes of recognized brands and were used in excellent state of preservation. Their liquid compositions were analyzed through GC/FID/MS following the same sample preparation, procedure, and analysis conditions as before. In terms of identification of chemical components, the only difference here was that a new lower-limit value for peak area was considered for the FID data equal to 2000 counts. In this way, only components with an area below that value were excluded. The obtained Perfumery Radars for unisex perfumes (U1–U6) are presented in Fig. 4.13 while for perfumes (U7–U11) are shown in Fig. 4.14.

First of all, it can be observed that, as for the feminine perfumes studied before in this Chapter, there are several discrepancies in the classifications obtained for the same perfume. It is also seen that certain olfactive families are typical or characteristic of the unisex fragrances like citrus, woody, and fresh. This is in agreement with the evaluation of Zarzo and Stanton as shown in Table 4.4.

Moreover, it should be noted that for some of these commercial perfumes there is not unanimity on the classifications as to their unisex (U) character, as shown in Table 4.6. With the exception of Voyage d'Hermès (Hermès), all the other perfumes present at least one

Table 4.6. Commercial Name, Brand, and Family Classification of the Selected Unisex Perfumes

No.	Perfume	Brand	Year	Osmoz (Firmenich)	ScentDirect	H&R	The Fragrance Foundation	LT&TS	Perfume Intelligence	ISIPCA	This Work
U1	Aqua Allegoria Herba Fresca	Guerlain	1999	–	–	–	–	Weird mint	Summer garden (U)	Aromatic-Citrus (U)	Green-Citrus
U2	Aqua Allegoria Pamplelune	Guerlain	1999	Citrus-Aromatic (F)	–	–	–	Floral-Glowing grapefruit	Citrus Rich (U)	Citrus (U)	Floral-Fruity-Citrus
U3	Bulgari Extrême	Bulgari	1996	Woody-Aromatic (M)	–	Fresh-Green (F); Fresh-Citrus (M)	–	Woody-Spicy (M)	–	Floral-Woody-Citrus (U)	Floral-Green
U4	Eau Parfumée	Bulgari	1992	Citrus-Aromatic (U)	Chypre-Fresh (F)	Chypre-Fresh (F)	–	–	Fresh-Citrus (U)	Floral-Woody-Citrus (F)	Musk-Woody-Floral
U5	CK Be	Calvin Klein	1998	Floral-Woody-Musk (U)	–	Fougère-Fresh-Woody (M)	Aromatic-Fougère-Fresh (U)	Fougère	Fresh-Woody-Oriental (U)	Floral-Musky (U)	Floral-Musk-Fruity
U6	Cologne	Thierry Mugler	2001	Citrus-Aromatic (U)	–	–	–	Steam clean-Floral (M)	Citrus-Fruity-Musk (U)	Musky-Citrus (U)	Musk-Citrus
U7	Gaultier 2	J. Paul Gaultier	2005	Oriental-Vanilla (F)	–	–	Woody-Oriental-Classical (U)	Musk-Floral-Green	–	Floriental (F)	Green-Floral
U8	Eau de Cartier	Cartier	2001	Citrus-Aromatic (U)	Floral-Fresh (F)	Chypre-Fresh (F)	Citrus Rich (U)	Violet leaf-Woody-Citrus	–	Floral-Woody-Citrus (U)	Oriental-Citrus-Fruity
U9	Voyage d'Hermès	Hermès	2010	Woody-Floral-Musk (U)	–	–	Woody-Fresh-Citrus-Fruity (U)	–	Fresh-Musk-Woody (U)	–	Woody-Fruity
U10	Eau de Gentiane Blanch	Hermès	2009	Woody-Floral-Musk (M)	–	–	Woody-Crisp-Green (U)	–	(U)	–	Woody-Musk
U11	Eau de Campagne	Sisley	1976	Floral-Green (F)	–	–	Green-Classical (U)	–	(U)	–	Floral-Citrus-Green

Their gender from each fragrance house or expert is also presented: Feminine (F), Masculine (M), or Unisex (U) fragrances.
Source: Adapted with permission from Teixeira (2011). © 2011, M.A. Teixeira.

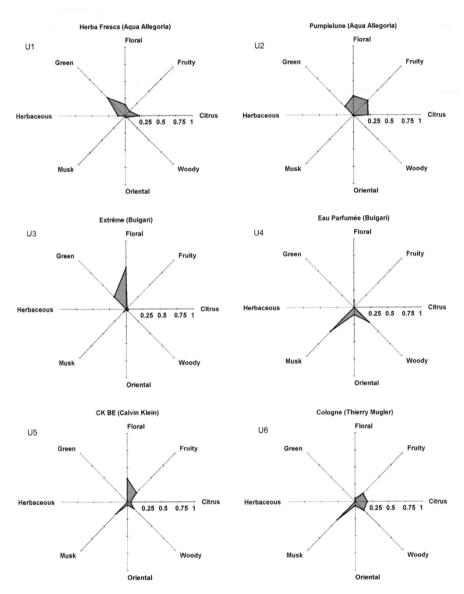

Fig. 4.13. PRs for unisex perfumes U1–U6. Adapted with permission from Teixeira (2011). © 2011, M.A. Teixeira.

classification as feminine (F) or masculine (M) fragrance. This stresses what was aforementioned, so that not only in the character classification but also in the gender assignment there are some differences.

From the comparison of the PRs predicted in this work and the classifications shown on Table 4.6, it is possible to see a very good

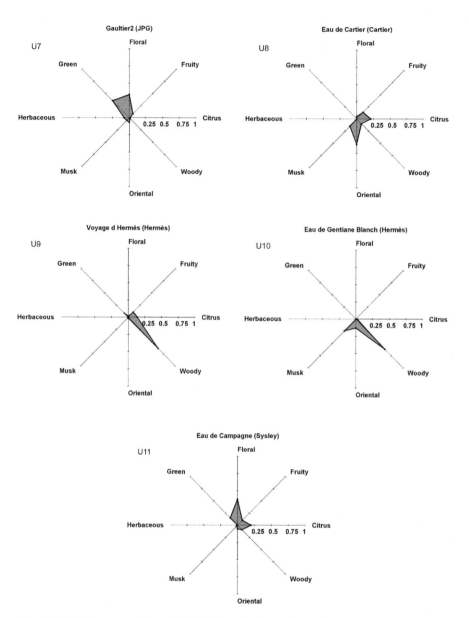

Fig. 4.14. PRs for unisex perfumes U7–U11. Adapted with permission from Teixeira (2011). © 2011, M.A. Teixeira.

agreement. For the great majority of the commercial perfumes, the PR methodology predicts the primary olfactive family, with few exceptions for Eau Parfumée (U4) and Eau de Cartier (U8) for which only secondary families were well predicted. Nevertheless, for perfumes like

U9, U10, or U11 predictions were obtained that match perfectly well all the classifications from the experts.

As a curiosity, it should be noted for these unisex perfumes the presence of the musk family, which gives sweetness to the perceived scent and which was not that common on the feminine perfumes studied before.

4.4.3 Evaluation of Odor Intensities of Similar Perfumes

As stated in the beginning of this chapter, perfumes can be classified in terms of their chemical composition (depending on the desired type of application) or their olfactory families (for character or quality analysis). Here, we will compare both the perceived odor intensity and character of similar perfumes (same brand and model) that were designed to be fine fragrances (e.g., *Eau de Toilette*) or used as lighter fragrances (e.g., *Eau Fraîche* or *After Shave*). For that purpose, two fragrances were selected: one feminine which was previously evaluated in terms of odor character only—Addict (Dior)—with its two versions, *Eau de Toilette* and *Eau Fraîche*, and one masculine fragrance—Euphoria Intense (Calvin Klein)—with the *Eau de Toilette* and *After Shave* versions. A comparison between the PRs is shown in Fig. 4.15.

The PR methodology was applied to these different fragrance versions which, as we know, will differ in their composition, especially in terms of alcohol and water content. First of all, a comparison between the compositions of these fragrances has shown that although they have

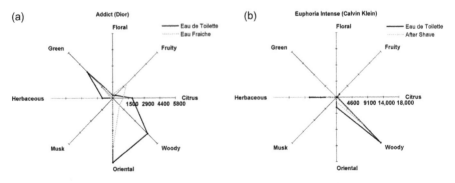

Fig. 4.15. Comparison between the odor intensities of the olfactive families of two different fragrances: (a) Eau de Toilette *versus* Eau Fraîche *versions of the commercial perfume Addict (Dior) and (b)* Eau de Toilette *and* After Shave *versions of Euphoria Intense (Calvin Klein). Adapted with permission from Teixeira (2011).* © 2011, M.A. Teixeira.

almost the same fragrant components within the mixture, some of them are present in significantly different concentrations, despite the obvious differences in ethanol and water content. For example, in Addict, limonene and linalyl acetate are, respectively, three and five times more concentrated in the *Eau de Toilette* than in the *Eau Fraîche*, while they have eight and four times more concentration in the Euphoria Intense *Eau de Toilette* than in the *After Shave*.

From the analysis of Fig. 4.15, it is clear that there are differences in both intensity and character between the two versions of Addict tested here (left) while for Euphoria Intense (right) the perceived character is similar though with little differences in the intensities of the olfactive families. Moreover, in terms of odor intensity, it is seen that the OVs for the dominant notes of each perfume are much higher in the woody note for Euphoria Intense (max $OV_j = 18,000$) than in the oriental note for Addict (max $OV_j = 5,800$).

In terms of character, Euphoria Intense is classified by experts as woody-aromatic by Osmoz (Firmenich, 2012) and ISIPCA (ISIPCA, 2010) and as woody-oriental by The Fragrance Foundation (Edwards, 2009) and the Perfume Intelligence (Perfume Intelligence, 2011). In this work, it is predicted as woody with powerful herbaceous and oriental *nuances*, thus showing good agreement.

4.5 EXPERIMENTAL VALIDATION OF THE PR METHODOLOGY

Hitherto we have explored some of the potential applications of the PR methodology, assessing its predictive performance by comparison with perfumers' olfactory ratings. Yet, another experimental validation of the PR methodology can be attained obtaining experimental PRs: experimental OV can be calculated from concentrations measured in the headspace of some commercial perfumes. Their comparison with the corresponding PRs predicted from the liquid composition of the perfumes will allow assessing the predictive capability of the PR methodology.

In such experiments, the headspace composition of three commercial perfumes belonging to different olfactive families was first analyzed by GC/FID/MS. Once their vapor composition was known, the corresponding PR was developed based on real concentrations in the gas phase. In Fig. 4.16, a comparison between the PRs predicted from the liquid composition and those obtained from HS-GC measurements is presented.

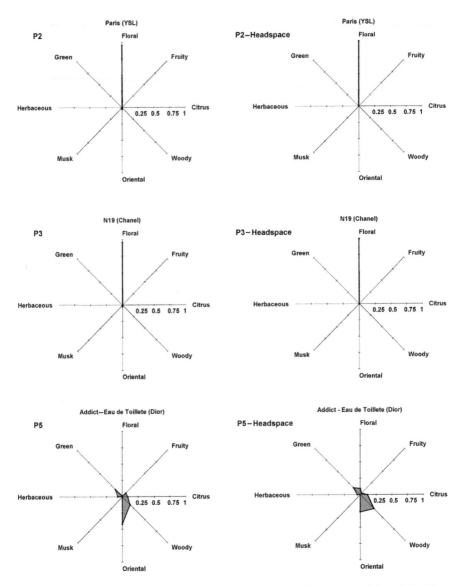

Fig. 4.16. Comparison between predicted PRs from the liquid composition (left column) and those obtained from headspace GC-FID-MS measurements (right column). Adapted with permission from Teixeira et al. (2010). © 2010, American Chemical Society.

From Fig. 4.16, it is possible to retain the great similarity between the experimental PRs and those previously obtained for perfumes P2, P3, and P5. Although for the first two perfumes (P2 and P3) there is a complete match between the predicted and experimental radars, for P5 the main olfactive family are the same, showing, once more, an

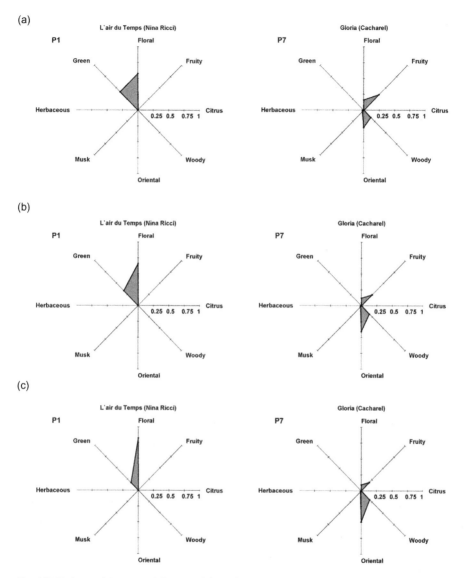

Fig. 4.17. Evolution of the perceived character of the perfumes P1 and P7 using the PR methodology with evapo-ration and diffusion over time: (a) t = 0 s; *(b)* t = 30 s; *(c)* t = 60 s. *Adapted with permission from Teixeira et al. (2010). © 2010, American Chemical Society.*

oriental-woody-green character. Only the fruity *nuance*, which was pre-dicted before, was not detected in the experimental radar. That being said, the experimental headspace classification shows a great similarity with the predicted PRs. Furthermore, throughout the identification of components in the headspace analyses, it was possible to see that the most powerful fragrances that were detected by GC/FID/MS were also predicted by the PR methodology. This experimental validation of the PR methodology has another great implication: it shows that the UNIFAC method is suitable for the prediction of the VLE of multi-component mixtures. This is mainly true when a qualitative evaluation is being performed (as previously discussed in Chapters 2 and 3 on the topic of the relevance of activity coefficients in multicomponent mixtures), although some deviations may occur.

Here, a parenthesis is made to highlight the importance of such parameter in the PR methodology. It could be argued to make a simplification and consider an ideal liquid solution. However, when comparing the results obtained for both cases, despite the fact that in some of them that qualitative evaluation may smooth the differences, in the majority of the perfumes large discrepancies are found in their classification. In this way, considering idealities for the liquid perfume mixture would lead to erroneous classifications of perfumes.

Nonetheless, it should be highlighted that headspace analysis is much more limited than it is for the liquid perfume because of the dilution ratio. Vapor concentrations are presented highly diluted in air, turning it more difficult to detect.

4.6 EVOLUTION OF THE PR WITH EVAPORATION AND DIFFUSION

So far, we have presented the PR as a tool for the classification of commercial perfumes into olfactory families. These radars represent the main olfactory families as if they were perceived at equilibrium concentrations (e.g., near the liquid–gas interface). However, when perfumers evaluate the olfactory space of perfumes, they often place the liquid in a paper blotter, fan it, and allow some time prior to the olfactory evaluation. This allows a faster evaporation of the most volatile species present in the mixture (mainly the ethanol). Because the PR methodology did not contemplate this effect, we decided to develop PRs including diffusion effects that take place over time. In this way, the diffusion model previously presented in Chapter 3 was applied in the PR methodology to account for the diffusion of the fragrance ingredients of each commercial perfume in air over time (t) and distance (z) (Teixeira et al., 2009). This diffusion model, as we said before, is based on Fick's law for diffusion and because it uses estimated data (diffusion coefficients, physicochemical properties) from the literature, it allows the prediction of the perceived odor over time and distance. Ahead, we present the application of the diffusion model to two feminine perfumes (P1—L'air du Temps from Nina Ricci and P7—Gloria from Cacharel) that will allow predicting the evolution of the odor character shown in the PR with time.

A comparison between the PRs at equilibrium conditions and after 30 and 60 s of evaporation and diffusion are presented in Fig. 4.17 for perfumes P1 and P7. It is clear in these two examples that when diffusion is taken into account, the concentrations in the gas phase above the perfume change with time as the perfume evaporates. Thus, the shape of its PR changes with time.

In the case of P1, it is seen that the initial shared dominant families, floral and green, have changed to a more intense floral character over time, although the green *nuance* is still perceived. On the other hand, for P7 the fruity and oriental dominant families have evolved to a stronger oriental-woody character with time. Strikingly, the evolution of the PR over time tends to get closer to the perfumers classifications. For L'air du Temps we predicted a floral-green character, although the perfumers classifications (Table 4.4) were mainly floral, except that of LT&TS which completely agreed with our PR. But if we consider the diffusion of the perfume, its radar becomes much more floral than green, matching what experts described before. For the perfume Gloria (P7), we see that its initial fruity-oriental character changed over time to a more predominant oriental-woody which is, in fact, in better agreement with the evaluations of perfumers. Thus, the incorporation of the diffusion model in the PR methodology tends to mimic the experimental procedure used for perfume classification and so it predicts more accurately the olfactive families.

4.7 FINAL CONSIDERATIONS ON THE PR METHODOLOGY

Although the application of the PR methodology for the classification of commercial perfumes using olfactive families has proven to give accurate predictions, the following remarks should be taken into account. It is important to highlight that some differences between the simulations and the classifications based on the olfactory perception of experienced perfumers may have to do with some factors discussed ahead. First, the approximation made for the water content in the evaluated perfumes: once we have used gas chromatographic analysis in a GC/FID/MS, water is not detected (FID is not suitable for detection of water molecules and MS operates only at low water concentrations, nonmoisture mode). Consequently, the compositions of the commercial perfumes had to be normalized, depending on the type of perfume (*Eau de Toilette* or *Eau Fraîche*) and according to the literature (Teixeira et al., 2010). Moreover, as previously discussed, a perfume may contain dozens of fragrance ingredients of different types in its composition. As a consequence, our methodology accounts for a large number of those and, thus, two issues have to be highlighted: (i) the assumption that chromatographic areas can be used to calculate the composition of the perfume (by assuming the same factor of response in the FID for all chemicals); (ii) the evaluation of the OV for each component is dependent on the availability of physicochemical (P_i^{sat}, M_i, UNIFAC interaction parameters) and psychophysical (ODT_i) data. For this last reason in particular, throughout this work, some other perfumes were discarded due to the limitation of available threshold data or the impossibility of attributing group parameters (UNIFAC) for fragrant molecules present in those perfumes.

Nonetheless, our PR is a modular methodology, since it combines different models for odor perception which can be updated or simply switched by others. In the case presented here, the UNIFAC method was used for VLE, Fick's law for diffusion, the OV for odor intensity, and the Family intensity model for odor quality. Yet, each of these stages can be modeled and/or complemented using different theories, models, or equations.

Despite the indisputable value of the PR methodology, it is expected that perfumers will continue to stick to their personal framework and F&F companies will continue to lay on their experience to develop new perfumes. However, the PR methodology can be adapted to the preferences of the perfumer or company. One of the most discrepant topics may be the classification of fragrance raw materials into olfactive families because each F&F company has its own. The number and type of olfactive families or their placement in the radar are also sensitive topics for each fragrance house. Nevertheless, the application of the PR methodology, even tailored as said, presents two major advantages: it is a predictive tool that can be applied in the preformulation stage of the perfume, so there is no need for experimental evaluation, and would provide a standard basis for comparison of different perfumes. Thus, we consider it would contribute to the *a priori* product design of fragrances, and would allow reducing cost and time of production.

4.8 CONCLUSION

The PR methodology is a perfume classification tool that is able to predict the primary olfactive family of essential oils and commercial perfumes of different types or gender. Within its modular structure it uses scientific models for predicting VLE (and diffusion as well) of fragrances, combined with psychophysical models for the qualitative classification of perfumes. Consequently, it reduces the arbitrariness of these classifications to the experimental evaluation of olfactory descriptors for pure fragrance chemicals, instead of relying exclusively on the sensorial perception of perfumers for the mixture. The PR presents itself as a valuable technique for the design of new and improved perfumed products.

REFERENCES

Amoore, J. E., Forrester, L. J., Specific Anosmia to Trimethylamine: The Fishy Primary Odor, *Journal of Chemical Ecology*, 2 (1): 49–56, 1976.

Amoore, J. E., Hautala, E., Odor as an Aid to Chemical Safety: Odor Thresholds Compared with Threshold Limit Values and Volatilities for 214 Industrial Chemicals in Air and Water Dilution, *Journal of Applied Toxicology*, 3 (6): 272–290, 1983.

Amoore, J. E., Forrester, L. J., Buttery, R. G., Specific Anosmia to 1-Pyrroline: The Spermous Primary Odor, *Journal of Chemical Ecology*, I (3): 299–310, 1975.

Arctander, S., Perfume and Flavor Chemicals, Montclair, NJ, Aroma Chemicals, 1969.

Artiga, M., Milla, D., The Evaluation of Beer Aging. *in Preedy, V., editor: Beer in Health and Disease Prevention*, London, Academic Press–Elsevier, 2009.

Avon, AVON Company, from http://knowledgeonline.youravon.com/content/beautyboost/2008B24/0808_quick_guide.pdf, Last Accessed on October 2012.

Axel, R., Scents and Sensibility: A Molecular Logic of Olfactory Perception (Nobel Lecture), *Angewandte Chemie-International Edition*, 44 (38): 6110–6127, 2005.

Bamforth, C. W., Russell, I., Stewart, G., Handbook of Alcoholic Beverages Series–Beer: A quality Perspective, Academic Press, 2008.

Brechbill, G. O., A Reference Book on Fragrance Ingredients, Hackensack, NJ, Creative Endeavor Books, 2006.

Buck, L. B., Unraveling the Sense of Smell (Nobel Lecture), *Angewandte Chemie-International Edition*, 44 (38): 6128–6140, 2005.

Buck, L., Axel, R., A Novel Multigene Family May Encode Odorant Receptors: A Molecular Basis for Odor Recognition, *Cell*, 65: 175–187, 1991.

Buettner, A., Schieberle, P., Evaluation of Aroma Differences Between Hand-Squeezed Juices from Valencia Late and Navel Oranges by Quantitation of Key Odorants and Flavor Reconstitution Experiments, *Journal of Agricultural and Food Chemistry*, 49 (5): 2387–2394, 2001.

Butler, H., Poucher's Perfumes, Cosmetics and Soaps, Boston, MA, Kluwer Academic Publishers, 2000.

Cain, W. S., Schiet, F. T., Olsson, M. J., de Wijk, R. A., Comparison of Models of Odor Interaction, *Chemical Senses*, 20: 625–637, 1995.

Caldeira, I., Belchior, A. P., Climaco, M. C., de Sousa, R. B., Aroma Profile of Portuguese Brandies Aged in Chestnut and Oak Woods, *Analytica Chimica Acta*, 458 (1): 55–62, 2002.

Calkin, R., Jellinek, S., Perfumery: Practice and Principles, New York, NY, John Wiley & Sons, 1994.

Callegari, P., Rouault, J., Laffort, P., Olfactory Quality: From Descriptor Profiles to Similarities, *Chemical Senses*, 22 (1): 1–8, 1997.

Chastrette, M., Trends in Structure-Odor Relationships, *SAR and QSAR in Environmental Research*, 6: 215–254, 1997.

Chastrette, M., Data Management in Olfaction Studies, *SAR and QSAR in Environmental Research*, 8 (3–4): 157–181, 1998.

Chastrette, M., Classification of Odors and Structure-Odor Relationships. *in Rouby, C., Schaal, B., Dubois, D., Gervais, R., Holley, A., editors: Olfaction, Taste, and Cognition*, UK, Cambridge University Press, 2002.

Chastrette, M., Desaintlaumer, J. Y., Sauvegrain, P., Analysis of a System of Description of Odors by Means of 4 Different Multivariate Statistical-Methods, *Chemical Senses*, 16 (1): 81–93, 1991.

Crocker, E. C., Henderson, L. F., Analysis and Classification of Odors: An Effort to Develop a Workable Method, *American Perfumer and Essential Oil Review*, 22: 325–356, 1927.

Distel, H., Ayabe-Kanamura, S., Martinez-Gomez, M., Schicker, I., Kobayakawa, T., Saito, S., Hudson, R., Perception of Everyday Odors—Correlation Between Intensity, Familiarity and Strength of Hedonic Judgement, *Chemical Senses*, 24 (2): 191–199, 1999.

Donna, L., Fragrance Perception: Is Everything Relative? *Perfumer & Flavorist*, 34 (12): 26–35, 2009.

Dravnieks, A., Current Status of Odor Theories, *Advances in Chemistry Series*, 56: 29–52, 1966.

DROM, DROM Fragrance Circle, from http://www.drom.com/, Last Accessed on 2011.

Edwards, M., The Fragrance Wheel, from http://www.fragrancedirectory.info/usadirectory/, Last Accessed on May 2011.

Edwards, M., The Fragrance Wheel, from http://www.fragrancedirectory.info/usadirectory/, Last Accessed on October 2012.

Ellena, J. C., Des Odeurs et des Mots, *Parf. Cosmet. Aromes*, 76: 63–64, 1987.

Ferreira, V., Lopez, R., Cacho, J. F., Quantitative Determination of the Odorants of Young Red Wines from Different Grape Varieties, *Journal of the Science of Food and Agriculture*, 80 (11): 1659–1667, 2000.

Firmenich, Osmoz Encyclopedia, from http://www.osmoz.com/, Last Accessed on May 2012.

Fisher, C., Scott, T. R., Food Flavours: Biology and Chemistry, Cambridge, UK, The Royal Society of Chemistry, 1997.

Gilbert, A., What the Nose Knows—The Science of Scent in Everyday Life, New York, NY, Crown Publishers, 2008.

Gottfried, J. A., Perceptual and Neural Plasticity of Odor Quality Coding in the Human Brain, Chemosensory Perception, Springer, 2007.

H&R, Haarmann & Reimer Fragrance Guide: Feminine Notes, Masculine Notes, Holzminden, Germany, 2002.

Haddad, R., Lapid, H., Harel, D., Sobel, N., Measuring Smells, *Current Opinion in Neurobiology*, 18 (4): 438–444, 2008.

Hoshino, O., Kashimori, Y., Kambara, T., An Olfactory Recognition Model Based on Spatio-Temporal Encoding of Odor Quality in the Olfactory Bulb, *Biological Cybernetics*, 79: 109–120, 1998.

ISIPCA, Classification Officielle des Parfums et Terminologie, Versailles (France), 2010.

Jaubert, J. N., Tapiero, C., Dore, J. -C., The Field of Odors: Toward a Universal Language for Odor Relationships, *Perfumer & Flavorist*, 20 (May–June): 1–15, 1995.

Jinks, A., Laing, D. G., The Analysis of Odor Mixtures by Humans: Evidence for a Configurational Process, *Physiology & Behavior*, 72 (1–2): 51–63, 2001.

Kraft, P., Eichenberger, W., Conception, Characterization and Correlation of New Marine Odorants, *European Journal of Organic Chemistry*, 19: 3735–3743, 2003.

Kraft, P., Bajgrowicz, J. A., Denis, C., Frater, G., Odds and Trends: Recent Developments in the Chemistry of Odorants, *Angewandte Chemie-International Edition*, 39 (17): 2981–3010, 2000.

Laffort, P., Dravnieks, A., Several Models of Suprathreshold Quantitative Olfactory Interaction in Humans Applied to Binary, Ternary and Quaternary Mixtures, *Chemical Senses*, 7 (2): 153–174, 1982.

Laing, D. G., Natural Sniffing Gives Optimum Odour Perception for Humans, *Perception*, 12 (2): 99–117, 1983.

Laing, G. G., Optimum Perception of Odours by Humans, Report, North Ryde, Australia, CSIRO—Division of Food Research, 1987.

Lawless, H. T., Descriptive Analysis of Complex Odors: Reality, Model or Illusion? *Food Quality and Preference*, 10 (4–5): 325–332, 1999.

Leffingwell, J. C., Leffingwell, D., Odor Thresholds, from http://www.leffingwell.com/odorthre.htm, Last Accessed on May 2012.

Mamlouk, A. M., Martinetz, T., On the Dimensions of the Olfactory Perception Space, *Neurocomputing*, 58: 1019–1025, 2004.

Mamlouk, A. M., Chee-Ruiter, C., Hofmann, U. G., Bower, J. M., Quantifying Olfactory Perception: Mapping Olfactory Perception Space by Using Multidimensional Scaling and Self-Organizing Maps, *Neurocomputing*, 52-4: 591–597, 2003.

Mata, V. G., Gomes, P. B., Rodrigues, A. E., Science Behind Perfume Design, Second European Symposium on Product Technology (Product Design and Technology), Groningen, The Netherlands, 2004.

McGinley, M. A., McGinley, C. M., Mann, J., Olfactomatics: Applied Mathematics For Odor Testing, WEF Odor/VOC 2000 Specialty Conference Cincinnati, OH, St. Croix Sensory Inc./McGinley Associates, P.A., 2000.

Milotic, D., The Impact of Fragrance on Consumer Choice, *Journal of Consumer Behaviour*, 3 (2): 179–191, 2003.

Moyano, L., Zea, L., Moreno, J., Medina, M., Analytical Study of Aromatic Series in Sherry Wines Subjected to Biological Aging, *Journal of Agricultural and Food Chemistry*, 50 (25): 7356–7361, 2002.

Niessing, J., Friedrich, R. W., Olfactory Pattern Classification by Discrete Neuronal Network States, *Nature*, 465: 47–52, 2010.

Ohloff, G., Importance of Minor Components in Flavors and Fragrances, *Perfumer & Flovarist*, 3: 1978.

Olsson, M. J., Berglund, B., Odor–Intensity Interaction in Binary and Ternary Mixtures, *Perception & Psychophysics*, 53 (5): 475–482, 1993.

Oyamada, T., Kashimori, Y., Kambara, T., Hierarchical Classification of Odor Quality Based on Dynamical Property of Neural Network of Olfactory Cortex, *IEEE*, 1997.

Parliment, T., Solvent Extraction and Distillation Techniques. *in R. Marsili, editor: Flavor, Fragrance and Odor Analysis*, New York, NY, Marcel Dekker, 2002.

Perfume Intelligence, Perfume Intelligence Encyclopaedia, from http://www.perfumeintelligence. co.uk/, Last Accessed on 2011.

Pintore, M., Wechman, C., Sicard, G., Chastrette, M., Amaury, N., Chretien, J. R., Comparing the Information Content of Two Large Olfactory Databases, *Journal of Chemical Information and Modeling*, 46 (1): 32–38, 2006.

Poucher, W. A., A Classification of Odors and its Uses, *Journal of the Society of Cosmetic Chemists*: 81–95, 1955.

Pybus, D. H., Sell, C. S., Chemistry of Fragrances, RSC Paperbacks, 1999.

Ross, S., Harriman, A. E., A Preliminary Study of the Crocker–Henderson Odor-Classification System, *The American Journal of Psychology*, 62 (3): 399–404, 1949.

Rossiter, K. J., Structure-Odor Relationships, *Chemical Reviews*, 96 (8): 3201–3240, 1996.

Rouby, C., Schaal, B., Dubois, D., Gervais, R., Holley, A., Olfaction, Taste, and Cognition, UK, Cambridge University Press, 2002.

Rowe, D., Chemistry and Technology of Flavours and Fragrances, United States, Blackwell Publishing Ltd–CRC Press, 2005.

Saito, H., Chi, Q. Y., Zhuang, H. Y., Matsunami, H., Mainland, J. D., Odor Coding by a Mammalian Receptor Repertoire, *Science Signaling*, 2 (60): 2009.

ScentDirect, Fragrance Genealogy, from http://www.scentdirect.com, Last Accessed on October 2012.

Schreiber, W. L., Perfumes, Kirk–Othmer Encyclopedia of Chemical Technology, 18, New York, NY, John Wiley & Sons, 2005.

Sell, C., The Chemistry of Fragrances–From Perfumer to Consumer, Cambridge (UK), RSC Publishing, 2006.

Sigma–Aldrich, Fine Chemicals Company. Flavors and Fragrances 2003–2004 Catalog, Milwaukee, WI, 2003.

Société Française des Parfumeurs, Classification des Parfums, from < http://www.parfumeur-createur.com/article.php3?id_article = 68 >, Last Accessed on October 2012.

Spehr, M., Munger, S. D., Olfactory Receptors: G Protein-Coupled Receptors and Beyond, *Journal of Neurochemistry*, 109 (6): 1570–1583, 2009.

Stevens, S. S., On the Psychological Law, *The Psychological Review*, 64 (3): 153–181, 1957.

Sturm, W., Peters, K., Perfumes, in Ullmann's Encyclopedia of Industrial Chemistry, Weinheim, Wiley-VCH, DOI:10.1002/14356007.a19; 2005.

Surburg, H., Panten, J., Common Fragrance and Flavor Materials–Preparation, Properties and Uses, Weinheim, Wiley-VCH, 2006.

Teixeira, M. A., Perfume Performance and Classification: Perfumery Quaternary–Quinary Diagram (PQ2D®) and Perfumery Radar. Department of Chemical Engineering, Faculty of Engineering of University of Porto, PhD Thesis, 2011.

Teixeira, M. A., Rodríguez, O., Mata, V. G., Rodrigues, A. E., The Diffusion of Perfume Mixtures and Odor Performance, *Chemical Engineering Science*, 64: 2570–2589, 2009.

Teixeira, M. A., Rodríguez, O., Rodrigues, A. E., Perfumery Radar: A Predictive Tool for Perfume Family Classification, *Industrial & Engineering Chemistry Research*, 49: 11764–11777, 2010.

Teixeira, M. A., Rodríguez, O., Rodrigues, A. E., Chapter 1: Odor Detection & Perception: An Engineering Perspective. in J. A. Daniels, editor: *Advances in Environmental Research*, Volume 14, USA, NovaPublishers, 2011.

TFF, The Fragrance Foundation. Key Fragrance Findings from The Premium Market Report 2010/11, from http://www.fragrancefoundation.org.uk/market-research.htm#top10, Last Accessed on 2011.

The Good Scents Company, from http://www.thegoodscentscompany.com/, Last Accessed on May 2010.

Turin, L., A Spectroscopic Mechanism for Primary Olfactory Reception, *Chemical Senses*, 21 (6): 773–791, 1996.

Turin, L., A Method for the Calculation of Odor Character from Molecular Structure, *Journal of Theoretical Biology*, 216: 367–385, 2002.

Turin, L., Chapter 11: Rational Odorant Design. in D. J. Rowe, editor: *Chemistry and Technology of Flavours and Fragrances*, Blackwell, 2005.

Turin, L., Sanchez, T., Perfumes—The Guide, London, Viking Penguin, 2008.

Weiner, I. B., Handbook of Psychology, New Jersey, John Wiley & Sons, 2006.

Wells, F. V., Billot, M., Perfumery Technology: Art, Science, Industry, New York, NY, John Wiley & Sons, 1988.

Wilson, D. A., Pattern Separation and Completion in Olfaction, *International Symposium on Olfaction and Taste*, 1170: 306–312, 2009.

Wise, P. M., Olsson, M. J., Cain, W. S., Quantification of Odor Quality, *Chemical Senses*, 25 (4): 429–443, 2000.

Yabuki, M., Portman, K. L., Scott, D. J., Briand, L., Taylor, A. J., DyBOBS: A Dynamic Biomimetic Assay for Odorant-Binding to Odor-Binding Protein, *Chemosensory Perception*, 3 (2): 108–117, 2010.

Zarzo, M., Relevant Psychological Dimensions in the Perceptual Space of Perfumery Odors, *Food Quality and Preference*, 19 (3): 315–322, 2008.

Zarzo, M., Stanton, D. T., Understanding the Underlying Dimensions in Perfumers' Odor Perception Space as a Basis for Developing Meaningful Odor Maps, *Attention Perception & Psychophysics*, 71 (2): 225–247, 2009.

Zellner, D. A., McGarry, A., Mattern-McClory, R., Abreu, D., Masculinity/Femininity of Fine Fragrances Affects Color–Odor Correspondences: A Case for Cognitions Influencing Cross-Modal Correspondences, *Chemical Senses*, 33 (2): 211–222, 2008.

Zwaardemaker, H., The Sense of Smell, *Acta Oto-Laryngologica*, 11 (1): 3–15, 1927.

Looking Ahead

In the opening remarks of this book, several topics within olfactory perception, odor modeling, and engineering perfume design were highlighted. However, the vastness and complexity of the main topics involved in the field of odor perception together with their interdisciplinary nature makes any attempt to exhaust all matters remarkably difficult.

Even so, throughout this book, several issues concerning the formulation of fragrances, prediction of odor intensity and character classification, olfactory analysis, or the evaluation of the release and performance of perfumed products were addressed. It is now time to summarize the major conclusions withdrawn from this work, but since we have covered only a small part of the multitude of research topics in Perfume Engineering, some questions were left unanswered. A long journey is still ahead in the understanding and prediction of odor perception.

The first step we have taken in the field of fragrances was, in fact, on the extraction and characterization of essential oils from aromatic plants using different techniques. However, the main target of our research group was always toward the development of tools and scientific models that could represent and, ultimately, predict the perceived odor released from a liquid mixture of fragrance ingredients. Following this line of thought, we presented, in the beginning of this book, our proposed methodology for odor perception which is a combination of four main steps: (i) the release/evaporation of fragrances to air, (ii) the diffusion of fragrance molecules through the surrounding air, (iii) the detection and intensity perception of these odorants, and, finally, (iv) the recognition and classification of odorants' character or qualities. Using this approach, we have developed the Perfumery Ternary Diagram (PTD®) methodology. It allows representing the perceived odor character of ternary mixtures of fragrance ingredients (or quaternary mixtures if considering, for example, a solvent-free basis). We have shown its application to the study of the effect of the base note on simple perfume formulations and its experimental validation. From that point, we have extended this tool to the Perfumery Quaternary–Quinary Diagram (PQ2D®) which allows the introduction of an extra fragrance ingredient or a solvent in the mixtures. It also allows the application to pseudoquinary systems, like when two solvents (e.g., ethanol and water) are introduced in the formulation. We have demonstrated the applicability of the PQ2D® by studying the effect of different base notes and fixatives on quaternary and quinary perfume mixtures. Finally, we have extended this approach to octonary perfume mixtures by introducing restrictions in the perfume composition and representing projections of a higher-order system like that with two top, middle, and base notes as well as two solvents. One of the great advantages of this software tool relies on its modular structure which allows changing models for each of the four steps that constitute it.

Following our methodology for odor perception, it was also important to evaluate the release of fragrance mixtures. For that purpose, the vapor–liquid equilibria (VLE) of several fragrance systems were evaluated experimentally and predicted by group-contribution methods showing good agreement. After this evaporation process, fragrances will diffuse in the air, so that a perfume will be perceived differently over time and distance. In this propagation process, an important variable for product design and formulation called product performance should be highlighted. In the fragrance industry performance can be evaluated in terms of the different perceived intensity and character of a perfume with time and distance from the releasing source. We have modeled perfume performance using a simple perfume diffusion model based on Fick's Second Law which was developed and implemented in MATLAB, together with performance parameters commonly used by the industry. The performance of different fragrance mixtures was also evaluated showing the effect of ethanol and fixatives in the perfume formulation.

Another topic we have addressed in this book was on the classification of commercial perfumes, something that is performed by experts (perfumers) at the industry, though with large discrepancies observed among fragrance houses. Our methodology, called Perfumery Radar (PR), aims for a scientific and standardized classification of perfumes. It uses models for the prediction of the VLE and Psychophysics for a qualitative classification of perfumes, instead of relying on the sensorial perception of perfumers only. It was shown that the PR methodology correctly predicted the primary olfactive family of four essential oils and several commercial perfumes. Consequently, its application in the industry could give important guidelines for the design of perfumed products.

Having reviewed some of the highlights of this book, we will embrace a brief journey through some current "*hot topics*" within Perfume Engineering along with a perspective of the application of research and development to the industry.

5.1 UNRAVELING THE SENSE OF OLFACTION

There is a wide range of opportunities for research and development within fragrance design, formulation, and perception. These should also be of great interest for the fragrance industry. A great part of that stems from the complexity of the olfactory system and the lack of knowledge we have about this sense. The olfactory system is an incredible receiving and integrating mechanism that is continually collecting multiple (suprathreshold) odor sensations. It is also a powerful chemical sense that is closely linked to the brain's emotional center. Thus, it is not surprising that it can play a strong role on our attitudes, moods, or decisions. It can influence people's cardiac rhythm, make us salivate, or stir our memories of pleasant times in our lives. For all this, smell can also make us buy products, which adds an economically relevant perspective. In this way, it is undisputable the relevance of the olfactory system in our lives.

However, several system-level organizations of olfaction still remain unexplored and so the mechanisms behind olfactory detection and recognition are still far from being completely understood. For that purpose, the Holy Grail within this field, which consequently will have effects on others, will be the understanding of all the biochemical and neurological mechanisms behind odor detection, odor recognition, and odor intensity (Gazzaniga, 2004; Reed, 2004). From that point, it will be possible to understand how we detect, recognize, and associate words to odors at the molecular and neuronal levels. Only with that level of knowledge one may establish more complex and reliable models for olfaction, unless it is simply by serendipity (Sell, 2006). A big step was taken in 1991 with the discovery of the olfactory receptors family by Buck and Axel (1991) (Axel, 2005; Buck, 2005), who were later awarded the 2004 Nobel Prize in Physiology or Medicine.

Thus, it is difficult to point a way to find those answers, but it is likely that it will encompass molecular genetics and neuroscience studies. In the former, there is a large number of research groups around the globe who have been focusing their research on the study of the interactions between odorant binding proteins (OBP), their peers odorants, and olfactory receptors (Taylor et al., 2008; Yabuki et al., 2010). In the latter, it is also believed that neuroscience may help understanding the role of olfactory receptors in human olfaction and, thus, contribute to unravel the sense of smell (Saito et al., 2004, 2009; Fleischer et al., 2009). Despite all this, the discovery of the functioning mechanisms of such phenomena at molecular level will take long. Until then, small steps must be taken, in little discoveries, some of which are postulated ahead.

5.2 THE ROLE OF PERFUME ENGINEERING

It is our conviction that Perfume Engineering can also help (at least in part) for the extension of this knowledge but will definitely play an important role in the evaluation of molecular interactions in liquid solutions, fragrance propagation in air, and in predicting the odor intensity and character of multicomponent fragrances. It is known that the fragrance industry, as well as others related, deal with different types of products containing fragrances (e.g., household cleaners, toiletries, and detergents) and currently use some tools similar to those addressed in this book during the product development process. For that reason and considering that fragrance products often contain a large number of ingredients, it would be of great interest to continue extending the PQ2D® methodology for N component mixtures. Our software is perfectly suited for the calculation of the odor intensity and character of N fragrant components but their graphical representation for higher-order systems remains complex, if not even impossible (and so the solution goes through the use of tables of data).

Another relevant contribution that Perfume Engineering might bring to the way we perceive odors relies on the classification and prediction of the odor character of fragrance ingredients and mixtures. For that purpose, we highlight at this point two different topics that may have an impact on the near future:

1. Extension and further validation (with perfumers) of the Perfumery Radar (PR) for the classification of multicomponent complex mixtures into olfactive families (or classes, like masculine/feminine perfumes). There are several related topics that may be evaluated herein: (i) number and type of olfactive families; (ii) layout of the olfactive families in the radar (relationships among olfactive families); (iii) location of the typical fougère family (typically masculine) which is a combination of different scents/families; and (iv) development of a theoretically based weighing criteria for subfamilies. Another application of the PR methodology that could be explored would be its application to wines and other beverages (classification). As suggested in *The Economist* magazine when reporting on the PR methodology. *It might also find a use in other trades that require a good nose. A wine radar would settle any argument between oenophiles as to whether a slight whiff of soggy cardboard indicates that a $1000 bottle of claret has become corked* (Kaplan, 2010). Although there are differences in the composition of perfumes and wines, the PR methodology could be applied to this field with some modifications and adaptations.

2. Application of structure–odor relationships (SOR) to the prediction of the odor quality(ies) of fragrance raw materials or mixtures. Modeling of odors based on olfactophore models (Kraft et al., 2000; Bajgrowicz et al., 2003; Kraft and Eichenberger, 2003) is something already under development by fragrance companies. It has also been applied to different olfactory notes for the representation of molecular features that are responsible for a given odor (Bajgrowicz et al., 2003). Further details on these approaches can be obtained from review works in the field (Rossiter, 1996; Chastrette, 1997; Kraft et al., 2000; Sell, 2006).

Finally, we believe that Perfume Engineering has the potential for another application, namely in the field of olfactive marketing. This is a recent trend in the industry, a growing fashion in merchandising places although the power of scent and its influence on people is still not fully understood, as we have seen before. Companies typically have their brand marketing driven only by visual images or fashionable music (Moeran, 2007). However, there has been a shift in recent years toward esthetic or sensory branding. The idea remains the same: attract consumers by appealing to their senses (sight, hearing, or olfaction) and induce them to stop, smell, and buy their products. Having said that, why marketing of brands should not be oriented to olfactory sensations or emotions instead?

In fact, many companies have already started to associate pleasant scents to their brands, trying to make the scent recalling the brand to the consumer. To render this idea of sensorial marketing, there are plenty of examples with scenting applications: shops, malls, supermarkets, casinos, hotels, office buildings, cinemas, or department stores just to mention a few (Whiff Solutions, 2008; i-sensis, 2012; Scent Marketing Institute, 2012). As a result, if in 2007 the scent marketing industry was billed at $100 million, future predictions now point to reach up to $1 billion by 2015 (Bradford and Desrochers, 2009). Additionally, it was considered as one of the top 10 trends to be watched in the upcoming years (Thomaselli, 2006). Nonetheless, this effect of smell on consumer decision-making behavior, it is rather surprising that olfactory marketing still remains comparatively undeveloped as a discipline (Moeran, 2007). The method of application is typically by using air conditioning systems together with fragrance diffusing apparels, which are used for the diffusion of the fragrances in the surrounding environment. One important target would be to model this release and dispersion phenomena in closed environments, for which Chemical Engineering has plenty of tools. The principle for the modeling of odor perception should use the concepts of odor threshold concentration and perception of odor quality as its basis. Then the determination of the exact concentration of fragrance needed so that it is perceived by the customers as a pleasant odor around the room is a function of the diffusivities and those two previous parameters. For that, the modeling of such environments together with the dispersion of the odor using mathematical diffusion models or computational fluid dynamics (CFD) techniques would be of great value.

5.3 FRAGRANCE PERFORMANCE: DIFFERENT ROUTES

Another important topic inside the fragrance industry is related with the performance of their products (for which F&F companies have dedicated research groups). Fragrances are volatile compounds that do not often last long in air (especially if convection phenomena is involved), so that the performance of fragranced products is often measured by its pleasantness together with the lastingness (Gygax and Koch, 2001). However, lastingness can be improved by means of controlled release mechanisms or devices. There are a number of typical routes to develop long-lasting fragrances that can be related to the chemical synthesis of new odorant molecules, use of profragrances or microencapsulation techniques, among others. The increase of fragrance concentrations in air or any other media presents several advantages: it is not only socially enjoyable but it is also useful for applications like masking malodors. Whenever a persistent odor remains on the clothes, on our body, or in the air, the use of techniques for slow fragrance release and long-lasting scents is valuable (Stora et al., 2001; Herrmann, 2007).

In this way, one of the approaches relies on chemical synthesis of new odorant molecules (or simply by changing small parts of it, the so-called group substitution). For example, it is possible to reduce odorant volatility without changing its odor characteristics. These methods are already used at the industry, as for example with the Doremox® molecule: a derivative of rose oxide, chemically modified by increasing its molecular weight and resulting in a 10-fold less volatile molecule than its precursor (Watkins et al., 1993).

Another technique is the use of dynamic mixtures where fragrances are mixed together with fragrance precursors, called profragrances, and reversible covalent reactions occur (Levrand et al., 2006; Herrmann, 2007). In this process, the fragrance is chemically produced after a trigger is activated, and then it is released. Different triggers may be used: enzymes, light, temperature or pH change, hydrolysis, or dynamic equilibrium. This process is expected to produce a slow release of fragrance materials in the air, thus producing a long-lasting perceived odor sensation. This concept (which is widely used in the pharmaceutical industry) was introduced to functional perfumery by Firmenich in the mid 1990s (Gautschi et al., 2001). Back then, for textile applications they used lipases in their detergent formulations to hydrolyze esters of fragrant alcohols, which in turn had better affinities toward fabrics than other fragrance ingredients.

Nevertheless, the most common process for the controlled release of highly volatile odorants has its origin in drug release techniques: the encapsulation of odorants into matrices or specifically designed capsules. This technology often makes use of polymeric microcapsules which act as small containers of a liquid solution to be released from the inner core under controlled conditions to address a specific purpose. In this way, it slows down their release and improves the lastingness of the fragranced product (Kukovic and Knez, 1996; Soest, 2007; Rodrigues et al., 2008, 2009; Martins et al., 2009; Haefliger et al., 2010; Specos et al., 2010; Teixeira et al., 2011).

5.4 FRAGRANCE STABILITY

Closely related to this last topic, is the fact that the fragrance industry is "speeding up," or in other words, is producing different fragranced products in large numbers and in short times. Twenty-five years ago, less than a hundred perfumes were introduced each year. Considering the year 2011 alone, there were more than 500 new fragrances launched in the market (although many of these will ultimately not succeed and be discontinued). A reason for this exponential growth is because unlike 25 years ago, the life of a perfume today is relatively short. According to Lefingwell, *in many cases now a fragrance will be very popular for a year or two, then the consumers move on* (Davies, 2009). This "volatility" of preferences crosses continents and cultures, because it is a consequence of the evolution of our society, and so pushes the industry to produce new formulations faster. If the PQ2D® methodology may speed up the design process of new fragrances, there may be other limiting steps for the formulation of the final product like the evaluation of product stability or maceration time. For the former, stability issues rise most of the times when fragrances are added to functional products which have different bases or matrices that are likely to interact differently with the fragrances. Nevertheless, some restrictions are also taken into account for fine fragrance development. From a consumer point of view, fragrance products must be unvarying in terms of quality. For that purpose, the main issues include: (i) chemical interactions between fragrance ingredients or components of the product base (where pH plays an important role also);

(ii) compatibility with the container or package material (e.g., should avoid evaporation losses); (iii) the possibility of product discoloration (e.g., due to formation of organometallic complexes with iron or other metal ions, due to UV irradiation effects); (iv) compound oxidation (e.g., unsaturated terpenes are likely to undergo air oxidation); and (v) loss of perfume strength by evaporation that leads to a change in odor character (Poucher, 1955; Calkin and Jellinek, 1994). The problem, however, is that although the majority of the reactions leading to perfume instability are known and understood (and so its behavior could be predicted *a priori*), in practice the combination of all these factors may turn it difficult to foresee. On the other hand, the maceration time for a perfume is usually long (several weeks or even months), in order to guarantee that (eventual) reactive processes are controlled and that the product is kept stable ("steady state"). The problem here is that the maceration period often exceeds the time that would be really necessary to obtain the desired organoleptic properties for the perfume (Lopez-Nogueroles et al., 2010). In this way, further studies that would encompass the prediction of the stability of fragrance products and optimize the maceration time needed would contribute to decrease the production time and economic costs.

5.5 INTEGRATIVE APPROACH FOR PRODUCT DEVELOPMENT: THE ROLE OF MICROECONOMICS AND PLANNING STRATEGIES

Recalling the classical perspective of Product Engineering presented in Chapter 1, it encloses the methodology following the quatriplete needs—ideas—selection—manufacture: it starts with the identification of consumer needs, which leads to the development of ideas and, then, their selection is made up to the manufacture of the product (Cussler and Moggridge, 2001; Wesselingh et al., 2007; Ulrich and Eppinger, 2011). However, apart from the similarities these authors share among them, there are several other approaches that have been proposed for chemical product development over the last decade. Recently, some valuable contributions have been added, especially dealing with microeconomics for a sustainable product development. Within that line of thought, appears the industrial framework called Stage-Gate™ Product Development Process (STPDP) which uses product design strategies based on decision analysis (Cooper, 2001; Seider et al., 2010) or the approach from Bagajewicz (Bagajewicz, 2007; Bagajewicz, et al., 2011), who applies microeconomics evaluations from the beginning of product design.

In this way, we consider that economic evaluations/predictions of product cost together with the potential success in the market (e.g., for a new fragranced product or a new commercial perfume to be developed) would be of great value. It is known that it is extremely difficult to evaluate consumer preferences for a perfume, and thus to predict its success. So far, the triumph of a perfume in the market has been put on the hands of the perfumers, the brand, or celebrity that signs the product and the money that is spent on marketing and publicity. Nevertheless, the development of a model or approach that would consider a preevaluation of consumer demand by introducing pricing and microeconomics models in order to define the best properties for the desired product would be an excellent task for further studies and also, without a doubt, for the industry as well.

In sum, there are plenty of opportunities and new applications for fragrance research in the future ahead. For that, the understanding of how do humans perceive odors and mixtures of them will be crucial for the success in many different fields. Nevertheless, odors will always make part of our lives, ruling somehow our behavior.

REFERENCES

Axel, R., Scents and Sensibility: A Molecular Logic of Olfactory Perception (Nobel Lecture), *Angewandte Chemie-International Edition*, 44 (38): 6110–6127, 2005.

Bagajewicz, M., On the Role of Microeconomics, Multi-scale Planning and Finances in Product Design, *AIChE Journal*, 53 (12): 3155–3170, 2007.

Bagajewicz, M., Hill, S., Robben, A., Lopez, H., Sanders, M., Sposato, E., Baade, C., Manora, S., Coradin, J. H., Product Design in Price-Competitive Markets: A Case Study of a Skin Moisturizing Lotion, *AIChE Journal*, 57 (1): 160–177, 2011.

Bajgrowicz, J. A., Berg-Schultz, K., Brunner, G., Substituted Hepta-1,6-dien-3-ones with Green/Fruity Odours Green/Galbanum Olfactophore Model, *Bioorganic & Medicinal Chemistry*, 11 (13): 2931–2946, 2003.

Bradford, K. D., Desrochers, D. M., The Use of Scents to Influence Consumers: The Sense of Using Scents to Make Cents, *Journal of Business Ethics*, 90: 141–153, 2009.

Buck, L., Axel, R., A Novel Multigene Family May Encode Odorant Receptors: A Molecular Basis for Odor Recognition, *Cell*, 65: 175–187, 1991.

Buck, L. B., Unraveling the Sense of Smell (Nobel Lecture), *Angewandte Chemie-International Edition*, 44 (38): 6128–6140, 2005.

Calkin, R., Jellinek, S., Perfumery: Practice and Principles, New York, NY, John Wiley & Sons, 1994.

Chastrette, M., Trends in Structure–Odor Relationships, *SAR and QSAR in Environmental Research*, 6: 215–254, 1997.

Cooper, R. G., Winning at New Products: Accelerating the Process from Idea to Launch, New York, NY, Perseus Publishing, 2001.

Cussler, E. C., Moggridge, G. D., Chemical Product Design, Cambridge, MA, Cambridge University Press, 2001.

Davies, E., The Sweet Scent of Success, *Chemistry World*, 40–44, 2009.

Fleischer, J., Breer, H., Strotmann, J., Mammalian Olfactory Receptors, *Frontiers in Cellular Neuroscience* 3: 9, 2009.

Gautschi, M., Bajgrowicz, J. A., Kraft, P., Fragrance Chemistry–Milestones and Perspectives, *Chimia*, 55 (5): 379–387, 2001.

Gazzaniga, M. S., The Cognitive Neurosciences, Cambridge, MA, The MIT Press, 2004.

Gygax, H., Koch, H., The Measurement of Odours, *Chimia*, 55 (5): 401–405, 2001.

Haefliger, O. P., Jeckelmann, N., Ouali, L., Leon, G., Real-Time Monitoring of Fragrance Release from Cotton Towels by Low Thermal Mass Gas Chromatography Using a Longitudinally Modulating Cryogenic System for Headspace Sampling and Injection, *Analytical Chemistry*, 82 (2): 729–737, 2010.

Herrmann, A., Controlled Release of Volatiles Under Mild Reaction Conditions: From Nature to Everyday Products, *Angewandte Chemie-International Edition*, 46 (31): 5836–5863, 2007.

i-sensis, São João da Madeira, Portugal, from <www.i-sensis.com>, Last Accessed on 2012.

Kaplan, M., Making Sense of Scents–A New Way to Describe a Fragrance. The Economist. Science and Technology, in <http://www.economist.com/node/17672786>. 2010.

Kraft, P., Bajgrowicz, J. A., Denis, C., Frater, G., Odds and Trends: Recent Developments in the Chemistry of Odorants, *Angewandte Chemie-International Edition*, 39 (17): 2981–3010, 2000.

Kraft, P., Eichenberger, W., Conception, Characterization and Correlation of New Marine Odorants, *European Journal of Organic Chemistry*, 19: 3735–3743, 2003.

Kukovic, M., Knez, E., Process for preparing carries saturated or coated with microencapsulated scents. WO 96/09114, 1996.

Levrand, B., Ruff, Y., Lehn, J. M., Herrmann, A., Controlled Release of Volatile Aldehydes and Ketones by Reversible Hydrazone Formation–"Classical" Profragrances Are Getting Dynamic, *Chemical Communications*, 28: 2965–2967, 2006.

Lopez-Nogueroles, M., Chisvert, A., Salvador, A., A Chromatochemometric Approach for Evaluating and Selecting the Perfume Maceration Time, *Journal of Chromatography A*, 1217 (18): 3150–3160, 2010.

Martins, I. M., Rodrigues, S. N., Barreiro, F., Rodrigues, A. E., Microencapsulation of Thyme Oil by Coacervation, *Journal of Microencapsulation*, 26 (8): 667–675, 2009.

Moeran, B., Marketing Scents and the Anthropology of Smell, *Social Anthropology*, 15 (2): 153–168, 2007.

Poucher, W. A., A classification of odors and its uses, *Journal Soc. Cosmet. Chem*, 6, (2): 81–95, 1955.

Reed, R. R., After the Holy Grail: Establishing a Molecular Basis for Mammalian Olfaction, *Cell*, 116 (2): 329–336, 2004.

Rodrigues, S. N., Fernandes, I., Martins, I. M., Mata, V. G., Barreiro, F., Rodrigues, A. I., Microencapsulation of Limonene for Textile Application, *Industrial & Engineering Chemistry Research*, 47 (12): 4142–4147, 2008.

Rodrigues, S. N., Martins, I. M., Fernandes, I. P., Gomes, P. B., Mata, V. G., Barreiro, M. F., Rodrigues, A. E., Scentfashion®: Microencapsulated Perfumes for Textile Application, *Chemical Engineering Journal*, 149: 463–472, 2009.

Rossiter, K. J., Structure–Odor Relationships, *Chemical Reviews*, 96 (8): 3201–3240, 1996.

Saito, H., Chi, Q. Y., Zhuang, H. Y., Matsunami, H., Mainland, J., Odor Coding by a Mammalian Receptor Repertoire, *Neuroscience Research*, 65: S76, 2009.

Saito, H., Kubota, M., Roberts, R. W., Chi, Q. Y., Matsunami, H., RTP Family Members Induce Functional Expression of Mammalian Odorant Receptors, *Cell*, 119 (5): 679–691, 2004.

Scent Marketing Institute, http://www.scentmarketing.org/., Last Accessed on October 2012.

Seider, W. D., Seader, J. D., Lewin, D. R., Widagdo, S., Product and Process Design Principles: Synthesis, Analysis and Evaluation, Asia, John Wiley & Sons, Inc., 2010.

Sell, C., The Chemistry of Fragrances–From Perfumer to Consumer, Cambridge, UK, RSC Publishing, 2006.

van Soest, J. J. G., Chapter 20: Encapsulation of Fragrances and Flavours: a Way to Control Odour and Aroma in Consumer Products, *in R. G. Berger, editor: Flavours and Fragrances: Chemistry, Bioprocessing and Sustainability*, Berlin Heidelberg, Springer-Verlag, 2007.

Specos, M. M. M., García, J. J., Tornesello, J., Marino, P., Vecchia, M. D., Tesoriero, M. V. D., Hermida, L. G., Microencapsulated Citronella Oil for Mosquito Repellent Finishing of Cotton Textiles, *Transactions of the Royal Society of Tropical Medicine and Hygiene*, 104: 653–658, 2010.

Stora, T., Eschera, S., Morris, A., The Physicochemical Basis of Perfume Performance in Consumer Products, *Chimia*, 55: 406–412, 2001.

Taylor, A. J., Cook, D. J., Scott, D. J., Role of Odorant Binding Proteins: Comparing Hypothetical Mechanisms with Experimental Data, *Chemosensory Perception*, 1 (2): 153–162, 2008.

Teixeira, M. A., Rodríguez, O., Rodrigues, S., Martins, I., Rodrigues, A. E., A Case Study of Product Engineering: Performance of Microencapsulated Perfumes on Textile Applications, *AIChE Journal*, 58 (6): 1939–1950, 2011.

Thomaselli, R., Trends to Watch in 2007, *Advertising Age*, 77 (51): 10, 2006.

Ulrich, K. T., Eppinger, S. D., Product Design and Development, New York, NY, McGraw-Hill, 2011.

Watkins, H., Liu, O. C., Krivda, J. A., Use of Tetra:hydro-4-methyl -2-phenyl-2H-pyran as a Perfume Ingredient–Capable of Imparting a Very Fresh and Green Odour Note with a Rosy Character. Firmenich SA, 1993.

Wesselingh, J. A., Kiil, S., Vigild, M. E., Design and Development of Biological, Chemical, Food and Pharmaceutical Products, Chichester, Wiley, 2007.

Whiff Solutions, LLC, www.whiffsolutions.com., Last Accessed on October 2012.

Yabuki, M., Portman, K. L., Scott, D. J., Briand, L., Taylor, A. J., DyBOBS: A Dynamic Biomimetic Assay for Odorant-Binding to Odor-Binding Protein, *Chemosensory Perception*, 3 (2): 108–117, 2010.

Printed and bound by CPI Group (UK) Ltd, Croydon, CR0 4YY

03/10/2024

01040413-0012